MATHEMATICAL GAMES, ABSTRACT GAMES

João Pedro Neto

Jorge Nuno Silva

Dover Publications, Inc.
Mineola, New York

The board diagrams in this book were made with the PostScript language from Adobe Systems and derived from the original set of board-design functions implemented by Cameron Browne.

Bibliographical Note

Mathematical Games, Abstract Games is a new work, first published
by Dover Publications, Inc., in 2013.

International Standard Book Number

ISBN-13: 978-0-486-49990-1
ISBN-10: 0-486-49990-1

Manufactured in the United States by Courier Corporation
49990101   2013
www.doverpublications.com

To my family
JPN

To Iguana, unreal llama
JNS

Contents

Chapter 1

The World of Games

Playing is an activity as old as civilization. There are two usual meanings for the word *play*: any childish activity without rules, and game playing, where rules are essential. This book is about games.

Some games are thousands of years old; they were probably the first strictly mental activities created by man. Nowadays, we learn about old games mainly when they are related to recreational mathematics. An example is Mancala, which uses a board reminiscent of an abacus, an old calculation device.

It is natural to classify games according to their rules. This book does not deal with games where chance plays a role (with the use of dice, for instance), or where there is some hidden information that a player has and his adversary does not (like most card games). Games that avoid these two situations are usually called abstract games.

We do not intend to write down the rules of chess or checkers. Our goal is to present a set of games, some not yet ten years old, that can help the reader and his family and friends find a leisure activity, especially those who enjoy intellectual challenges. Almost all the games treated here (the main exceptions are Go and Hex) have not yet been explored, but were chosen because the authors found in them strategical or tactical qualities, enough to offer hours of ludic pleasure.

We tried to produce a visually appealing and user friendly text, inviting the readers to play abstract games. Some more serious analysis is encouraged, for the enthusiasts.

When learning a game, it is strongly suggested that board and pieces be brought in place and the moves played out. Otherwise, it is easy to misunderstand some part of the rules or other aspects of the games. The authors are available for any clarification, mainly by e-mail.

This book is divided in the following parts:

1) This chapter, which consists of an introduction, an historical survey, and a superficial description of the actual panorama of board games.

2) A section dedicated to games for two players. For each of them, we describe the rules, the necessary material, and make some strategical and tactical commentaries on some positions.

3) A full chapter is occupied with the introduction of a class of mathematical games, which can be, at least in principle, completely analyzed in a quick and efficient manner. Nim and games on graphs are two such examples.

4) A section on games for three players. We present a set of games fit for the occasion when three people want to play and nobody wants to be left out. When the number of players is larger than two, new social complexities

arise, such as alliances, diplomacy, threats, and bluffs. A good game for these situations should be interesting without promoting personality clashes or extra-game arguments.

5) A glossary, to help minimize the ambiguity of the rules. It also explains the notations we use throughout the book for boards, positions and moves.

Brief Historical Journey

The expression "mathematical game" can be used to refer to a game, to a puzzle or to a problem of any degree of difficulty. The history of mathematics shows that mathematicians of all ages dedicated some of their energies to activities that could be classified as games. Some fields of mathematics were born this way.

All civilizations produce and play games. We do not know why, but their cultural and educational relevance is clear.

The games we'll be treating in this book are usually called *abstract* or *mathematical* games, but sometimes also *games of strategy*. We need some boards, that we'll describe, but in some cases pencil and paper, or piles of beans, are enough to play them.

The oldest game for which we know the rules is the Royal Game of Ur, from Babylon. It was found in the 1920s, and it is just a race between two opponents.

Two sets of pieces and a set of tetrahedrical dice can be seen here.

We think that the pieces used to move in the directions illustrated on the next page, being the length of each dispacement decided by the throwing of the dice.

The first player to finish his course would be the winner.

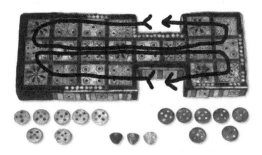

In ancient Egypt's tombs, as in the *Book of Dead*, we can find references to game playing, as can be seen in the picture below, from an inscription found in Nefertari's tomb:

or in this representation of players:

The two games depicted here are Senet (on the left) and Mehen (on the right, which has been identified with a representation of the god of games, the serpent god). Senet, which is also a race game, has a strong religious relevance. In the game represented above it looks like there is only one player. However, it is life after death that is being gambled on against the god of the afterlife.

Another race game from ancient Egypt was Dogs and Jackals:

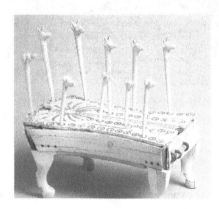

Once again we find almost independent courses. The players do not inter-act much during play. The lines connecting some pairs of holes (one "good" and one "bad"), to add drama to the game, are the first examples of such devices, well known in the popular Chutes & Ladders.

Nine Men Morris is a typical example of a family of alignment games. Probably known already in ancient times, only very recently (1996), through a very impressive computer analysis, it was found that, with perfect play, the game is a draw. It was very popular in western Europe in the Middle Ages.

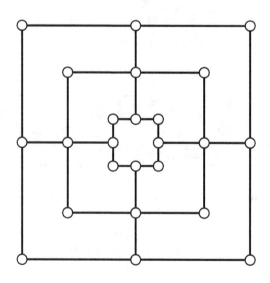

Ludus Latrunculorum, the soldier game, was a favorite of the Roman army. It is a military strategy game, the board representing a battlefield, the pieces representing soldiers. We are not sure about the original rules. The board was rectangular with variable dimensions. Some were carved in stone and can still be found in several Roman monuments.

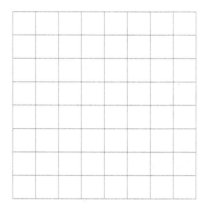

Alquerque, the grandfather of checkers, is also very old. We can find its vestiges in ancient monuments, raising difficult questions about its real age.

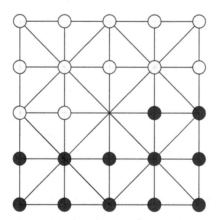

Archimedes (287-212 B.C.) described a geometric puzzle, the stomachion. It is a game similar to the tangram (a Chinese puzzle, still popular in the West), made up by fourteen plane shapes that can be assembled to make a square.

15

In the *Sand Reckoner*, Archimedes is believed to have described the combinatorial properties of stomachion, but that description has not survived.

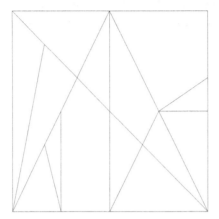

In the Middle Ages many board games were used. From the tenth century Ireland, the board of Fithcheall:

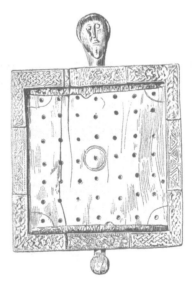

In this game, one of the two players takes the central square with his king, surrounding it with less valuable pieces. To win, this player needs to take his king to the periphery of the board. His adversary has more pieces, but no king. To win, this player needs to kill the king. The movements were orthogonal, like Chess rooks, and only custodian captures were permitted, surrounding the victim (see Glossary).

In the Middle Ages the erudite classes used to play several games. Some circulated only among universities, churches, and other educated places, where people could understand the complexity of the rules. One such example was Rithmomachia, also known as the Philosopher's Game, which was associated with the teaching of Boethius's *Arithmetic* for five hundred years.

Rithmomachia was a pedagogical game, especially designed to grasp some numerical relations, such as progressions.

Besides its practical utility, arithmetical knowledge had religious and moral value. Accordingly, Rithmomachia had an important role in the education of the learned classes for a few centuries.

The movements depended on the shape of the pieces, and captures depended also on the numbers each piece displayed. Victory was accomplished by occupying some squares in the adversary's half-board, with the corresponding numbers in special progressions.

Luca Pacioli, one of the leading mathematicians of the 15th century, was also a recreational mathematics fan. He wrote what is considered to be the first book on this subject. One of the puzzles described there is the now famous Chinese Rings:

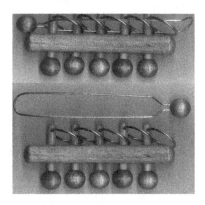

The Irish mathematician Sir William Rowan Hamilton, in 1857, created a game, Icosian, which was a commercial flop. However, it was related with a very important concept in graph theory, *Hamiltonian circuits*.

One of the oldest games from which chess developed is Chaturanga, a popular game in India during the 6th century (played by four people).

Chess rules have varied a lot since then. In the Middle Ages, piece movement was sometimes decided by the use of dice. From the 16th century on, chess became more popular than Rithmomachia among educated classes in Europe. Today chess is played by millions, and the Philosopher's Game is history.

18

Mathematical games are usually a favorite subject in books focused on the popularization of mathematics, and even in some pedagogical texts. However, math here is always apparent. In the games we call abstract the situation is different, the mathematical content is hidden. Playing the game is, in an abstract way, doing mathematics with the ludic side of each activity dominant.

G. H. Hardy, one of the most important mathematicians of the 20th century, said that the only difference between a chess problem and a mathematical theorem lies in their relevance. Abstract games and pure mathematics are the same... .

We believe that the practice of good games nurtures the intellect. We do not know how this mechanism actually works, but we believe that some good comes from playing interesting board games. Below we'll try to pinpoint the qualities that make a game worth playing.

The State of the Art

In some human activities, as in science and philosophy, today's production has attained such levels as to lead us to consider that we live in a golden age. The same is true for the market of games. In the past 25 years, board games have been invented at a pace never matched previously. In spite of the advent of computer games, new board games appear in quantity and quality, mainly in Germany, France and the USA. It is difficult, of course, to find a commercial success like Monopoly or Risk, but the German game from 1994, The Settlers of Catan, sold millions and is still selling well.

If we focus on abstract games, we find a different reality. There is some serious production of board games — Gigamic and Kris Burm's Gipf Project are good examples — but the Internet seems to be the natural habitat for this kind of game. Here, the communication among people sharing the same interests, but geographically apart, is easy. The emergence of dedicated forums, and other virtual platforms, to share information has triggered an outburst of creativity. Abstract games are particularly well adapted to this means of communication: it is easy to describe the rules and the boards, using only keyboard characters. It became possible to show a new game immediately after its conception to an interested public. In a few days a game can be tried, improved, or dismissed. This new version of postal chess, in which a turn could take weeks, is now at our fingertips. New servers were created to administer thousands of games.

Among traditional abstract games the best known are chess, with its regional dialects, Go, several Mancala games, Alquerque, and checkers.

During the 19th century and the first half of the 20th some modern classics were born, like Othello/Reversi, Chinese Checkers, Four-in-a-Row and Renju (essentially a five in-a-row with a complex protocol of initial rules). In our times, hundreds of new games appear during each decade. Most of them are just variations on old games, but others contain new ideas and interesting concepts.

Computer science plays a special role in the development of new games. It is not difficult to implement an algorithm that emulates a player. Some of the best chess programs play already at the grandmaster level. A landmark was Zillions of Games, which appeared in 1998. This application can implement a large quantity of different games (see the Electronic References).

Even without special software, it is possible to find someone on the Internet willing to try a new game. A few explorations later, it becomes clear whether the game has potential or if, on the contrary, it does not have any ludic qualities. When strong and weak points of a game are spotted, it is possible to improve the game, or to abandon it. The overall quality of board games increases this way. Possibly, if chess, Go, or checkers were invented today, they may not have the success they know. Other games, like the ones presented here, if created under different circumstances, could have been very successful. We hope that this book can help keep alive some of these games. They deserve to be passed on, if not for their history, then for their intrinsic value.

Families of Games

It is natural to classify games into families according to victory goal. If a game has more than one way of winning, the game belongs to more than one family naturally. We identify the following:

Territory games — A player tries to get as much area as he can (the calculation of the area depends on the rules of the particular game). In this book we present Go, Anchor, and Dispatch.

Blocking games — The winner is the player that blocks the adversary from any legal move. We include Amazons, Iqishiqi, Pawnographic chess, Campaign, Hobbes, UN, and all the Nim games.

Capture Games — The winner is the first to capture a set of pieces. In this book: Annuvin, Gogol, Nosferatu, and Hobbes.

Positional games — Victory depends on disposing one or more pieces in a certain part of the board. We have Aboyne, Pawnographic chess, Epaminondas, Gogol, Iqishiqi, and Slimetrail.

Pattern games — A pattern, usually a line, must be formed with the pieces. Examples: Gomoku, Havannah, Intersections, Semaphore, SanQi, and Stooges.

Connection games — The winner is the player that first forms a group of pieces satisfying some condition (for instance, connecting two sides of the board). This book presents a wide variety of connection games: Hex, Y, Nex, Gonnect, Jade, Havannah, and Lines of Action.

Inventing a new game is not easy, but the creation of a family has a different level of difficulty; here we need a completely new concept, not only a new set of rules. Besides, this new concept should be able to generate several good original games.

Variants

Everybody familiar with the game of checkers knows the *giveaway* variant, i.e., with the usual rules (capture is mandatory), the winner is the first player unable to move. Most of the traditional games have variants or other ways of playing, not as exact as the "official rules." Some of these variants have a regional character. For example Portuguese checkers and French checkers are different games, and English checkers is still another. Chess also has regional variants (actually, the international rules of FIDE are nothing more than a regional variant that went global), like Xiang-Qi (in China) and Shogi (in Japan). When there are several regional variants of the same game, the expression families of games is used (not with the previous meaning). Some examples: the family of chess, the family of checkers, and the family of Mancala (with hundreds of variants, mainly from Africa).

Traditional variants only account for a small part of the total. Most of them are modern. A good example is chess, which has thousands of variants, including many good games. Some of these were created by leading professional players, such as Laska (a checkers variant by Emanuel Lasker) and Fischer chess (a chess variant by Bobby Fischer).

How does one create a variant? Quite often a variant appears due to a lack of clear communication of the original rules. Probably, a new traditional version was created when a traveller was explained a game in a foreign language. Other variants were born due to deficient interpretation of written rules (which could happen with this book). But most variants were made on purpose. When a game is analyzed, played, and its strategy and tactics understood, then some changes start making sense. When the changes work well and give rise to an appealing game, a new variant is born.

The way new variants arise depends a lot on the particular game. But there are some general principles. There are meta-rules (also known as mutators) that change consistently a big class of games. For instance, the progression meta-rule, according to which the first player plays once, the second twice, the first player now plays three times, and so on. Usually this meta-rule is excessive, and has to be softened (see an example in Y). Applied to chess, with the restriction that "a sequence of moves ends before its natural length if a check occurs" gives rise to the well known variation progressive chess.

Here we present a list of other meta-rules:

a) Dagger (see Glossary): One of the players can play twice in one turn (see Gomoku, p. 62).

b) Introduction of neutral pieces (pieces that do not belong to either player), with the convention that each player can change two neutral pieces for two of his own and one of his by a neutral (see Nex, p. 98).

c) Pocket: Each player can pass his turn, saving that turn for a later moment (as if he had bought a dagger, the price being a pass).

d) Restrictions: To build an initial handicap to account for the unbalanced levels of the players. For example, in chess, the strongest player could start without the Queen's Knight and play White.

e) Pie rule (see Glossary): This could be the most important meta-rule. The first player does *n* moves (with both colors) and his adversary chooses the color he wants for the rest of the game. This rule makes it possible to balance some games where the first player has, without the pie rule, great advantage, such as Gomoku. This rule is assumed in several games in this book.

We can even combine meta-rules, if the particular game is a good fit for that. For example, progressive chess with a pie rule with $n = 2$ is a good possibility.

Quality control

How can we tell a good game from a bad one? How can we assess the quality of an abstract game? Which properties would we like to find in a game so we would spend time with it? There are no final answers to these questions, but we can point to some factors.

One of the most important elements is *depth*, or strategic complexity. How specialized can one be in one game? For example, tic-tac-toe has a very low complexity; it is easy to master its strategic subtleties and come out with a plan that leads consistently to draw games. On the opposite side is chess; there are several levels of sophistication at which chess can be played, from beginner to grandmaster.

Consider the following definition: A player X is one level above player Y if X beats regularly Y two out of three times. In chess, this accounts for a difference of about 100 ELO points (ELO is a system designed to assess the relative strength of the players.) If the beginner level corresponds to about 1,000 points and the best player ever has a level of 2,900, there are 19 levels of playing the game (actually a little more is true, since for stronger players the ELO classification does not grow so fast).

Tic-tac-toe may not have more than three levels (if we assume that children do not know the best strategy). Checkers is between tic-tac-toe and chess. Go (see p. 51), the old traditional game, has more than 30 levels of specialization. Check Thompson's article [THO] for an early discussion on these matters.

Clarity is another important issue. Clarity tries to answer the question "How difficult is it to create a good tactic or strategy?" The easier it is to visualize moves in the future, the greater clarity a game has. If a game has little clarity it becomes difficult to consider possible moves and anticipate the adversary's threats. It is nice when our victory is due to wise planning, not to a blunder of the opponent. Probably the clearest game in this book is Hex, and a very obscure one is Lines of Action.

Another important property is _drama._ A game has drama if it is possible to overcome a difficult situation by surprising strategic or tactical moves (for instance, sacrifices.) Chess is a great example of a dramatic game, as we can see in the problems in the literature. If a game between a weak player and stronger one is interrupted, and the weaker player's position given to a master, the game can still be interesting. Very deep games, like Go, usually have drama. But, on the other hand, a good game should be _decisive_: there must exist a situation towards which a player conducts the game that ensures him of the victory, independent of the level of his adversary. A game without this characteristic becomes confusing, producing cyclic dramatic events, with no end.

The average _time_ a game takes is also important. A chess game takes about 40 moves. One of checkers slightly less. Tic-tac-toe takes nine moves, at most. A game of Go can, after a 100 moves, be undecided. Of course it all depends on the free time each of us has, but nowadays games that take too long are penalized. (if Go were invented now, it would have a hard time surviving.) Games like tic-tac-toe, Hex, and Amazons, in which the cells of the boards are successively occupied, cannot go on too long, but a chess game could last, theoretically, over a thousand turns.

Ramification, that is, the number of possible moves a player can play, on average, in each move. It is in some ways the opposite of clarity. In

23

principle, the more possible moves there are, the less clear a game is. This property is important to computer scientists. The larger the ramification, the harder it is to design good software for a game. For example, chess's average ramification is about 40 moves, chess programs play very well. Go has an average ramification of 180 moves, the best computer programs play at beginner's level. Some of the games we present here have a ramification in the thousands, without penalizing clarity, as Nex or the progressive variant of Y.

The *interaction* is also important. This property addresses the level at which the pieces act on each other (a property introduced by Cameron Browne in [BRO]). Games with low interaction are just "individual race" games played at the same time, such as Chinese Checkers and Halma. In games with good interaction, it is possible to create complex configurations with the adversary's pieces, improving the quality of the game and increasing the number of relevant tactical moves. The use of neutral pieces plays a role here; see Sanqi, Nex, Iqishiqi and Hobbes.

We tried to include games that classify well according to these criteria. Most of them are very recent and almost unknown. Some are the result of our personal experience. We hope the readers enjoy the games.

Electronic References

The World of Abstract Games: www.di.fc.ul.pt/~jpn/gv A website of one of the authors, containing hundreds of board games from all ages and places. It is the main reference for the games for two players in this book.

Chess variants: www.chessvariants.com The ultimate website on chess variants.

Zillions of Games: www.zillionsofgames.com A commercial software package (there is a free demo version), that implements over a thousand board games, with an open specific language for the creation of new games.

Games by e-mail: www.gamerz.net/pbmserv/ An electronic games server created and administered by Richard Rognlie, where you can find opponents for dozens of games.

Board construction: http://www.di.fc.ul.pt/~jpn/gv/dobpt.htm Lots of boards used by the games described here are not easy to get. It is possible to print hard copies of boards and use chess or checker pieces or other materials to play. In this page we make available several PDF files that everyone can print. This page also contains hints and ideas to build physical game sets.

Associação Ludus: http://ludicum.org A Portuguese organization focused on Recreational Mathematics, including Mathematical Games.

João Pedro Neto
jpn@di.fc.ul.pt
http://www.di.fc.ul.pt/~jpn/

Jorge Nuno Silva
jnsilva@cal.berkeley.edu
http://jnsilva.ludicum.org/

Chapter 2

Games for Two

Aboyne

This game was invented by Paul Sijben in 1995. The name is an English verb invented by comic writers Douglas Adams and John Lloyd, meaning "to beat a master by playing so terribly that all known tactics are useless." (However, we do not advise this type of strategy.)

Materials

A hexagonal board with five hexagons per side, nine white stones, and nine black stones.

Initial setup:

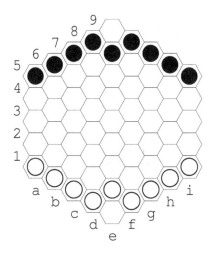

Definition

A piece is blocked if it is adjacent to a differently colored piece.

Rules

On each turn, each player must move a friendly unblocked piece. An unblocked piece may (i) move to an empty adjacent hexagon or (ii) jump a line of friendly pieces, where the jumping piece starts at one of the extremities of the line, and finishes at the other. If the other extremity is occupied by an enemy piece, this piece is captured. A white piece cannot move to e1. A black piece cannot move to e9.

Goal

Black wins by moving a friendly piece to e1, White wins by moving a friendly piece to e9. A player may also lose if all his pieces are blocked at the beginning of his turn.

Notes

In the next diagram, the pieces marked with an x are blocked. If it is White's turn, he can jump the white stone at g8 over the line of white stones, capturing the marked black stone at e6. Then, the initial white stone becomes blocked but releases f7 (and at the same time blocks d5). Next, Black can win the game by jumping c4 to e6 (capturing one white stone) because White's remaining stone becomes blocked.

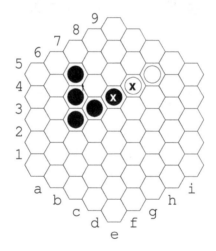

It is common that the moved stone blocks itself. This kind of move may be worth playing if it releases one or more friendly stones or reduces the number of possibilities of the player's opponent.

It is very important to evaluate outcomes of local battles and the potential blocks that a certain position may imply.

In the next diagram we see an endgame position. Black is almost winning, but how is he to achieve victory? Each player possesses only one unblocked piece.

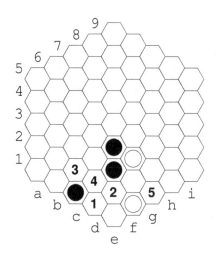

If the black stone moves to hexagon 1, White moves to hexagon 2, winning the game by blocking all black stones. (In fact, White also becomes blocked, but it is Black's turn.) Black must move to 3, forcing White to move to 5. (Any other choice blocks White.) Black moves to 4, and White cannot reply adequately. Then Black captures the white stone at f4, releasing the other two black stones. White's position is now hopeless.

Amazons

This game was invented in 1988 by Walter Zamkauskas. It has been studied in the domains of Artificial Intelligence and Combinatorial Game Theory. There is a yearly tournament among computer programs that has already achieved a quite advanced quality of play (e.g., the programs Arrow and Invader).

Materials

A square board with ten rows and ten columns, four black stones, four white stones, and 80 neutral pieces of a third color.

Initial setup:

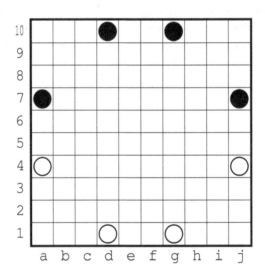

Rules

On each turn, each player executes two actions:

First, he moves a friendly Amazon. An Amazon slides, combining chess Rook and Bishop movements (orthogonally or diagonally) any number of empty squares in a straight line (like a chess Queen, but unable to capture).

Second, he drops a neutral piece on an empty square, provided that that square is within the moving range of the last moved Amazon (i.e., the Amazon could go to that square in one movement).

Goal

A player unable to execute his two actions loses the game.

Notes

Since one square gets occupied by a neutral on each turn, the game must end.

Here is an example of an early part of a game: White moves a friendly Amazon from g1 to d4 and drops a neutral piece at d9 which is valid because d4's Amazon could move to d9. Next, Black moves the Amazon at a7 to e7 and drops a neutral at i3. The board position is:

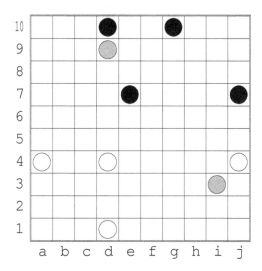

The games usually finish with a careful management of the remaining spaces. The player that ensures more empty squares available only to his own pieces wins the game. In the next position, it is White's turn. However, he lacks space to maneuver. He can move i7 (or i8) to h7 and place a neutral piece on the square where the movement started. Black can make any valid move and win because White is not able to conclude his next turn.

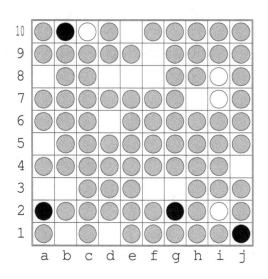

Reference

http://www.cs.ualberta.ca/~tegos/amazons/

Anchor

This game was invented by Steven Meyers in 2000. It is a territorial game reminiscent of Go (see p. 51), but on a board with two types of corners. The main novelty of Anchor lies in the redefinition of life and death concepts.

Materials

A hexagonal board with eight hexagons per side, approximately 100 white and black stones.

Initial setup:

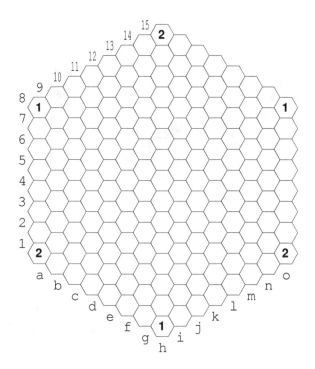

Definitions

Corners — Each player owns three corners. In the diagram, corners 1 belong to Black, corners 2 belong to White.

Group — A connected set of friendly pieces.

Anchor — A group is an anchor if it touches three or more board edges. A group touching two edges is an anchor if those edges do not meet in a corner that belongs to the adversary.

35

Life — A piece is alive if it belongs to an anchor; otherwise it is dead. In the next diagram, the pieces marked with "a" are dead.

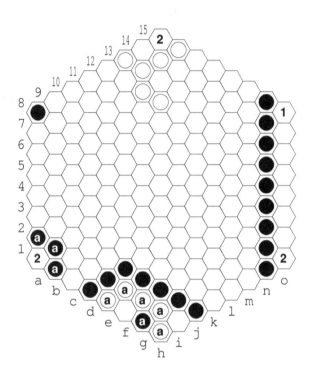

Rules

On the first turn, one player drops a black stone on an empty hexagon, and the other player decides which color he will play. White continues.

On each of the following turns, a player may pass or drop a friendly stone on an empty hexagon.

Goal

When both players pass consecutively, the game ends. Then each player removes any friendly dead pieces (called prisoners) and gives them to his adversary. Each prisoner is worth one point. Each empty hexagon surrounded by a player's stones (and eventually by the board edges) counts also as one point for that player. Each player counts his prisoners and surrounded hexagons. Whoever has the largest sum wins the game.

Notes

The winning condition is similar to Go's. The main difference consists in the evaluation of hexagonal territories.

The game's tension occurs in the constant necessity of cutting potential anchors of the enemy. Controlling the center is relevant (more so than in Go) because it allows a privileged access to connect several groups of pieces in order to make them alive by touching different board edges.

Creating anchors is important, but it is also essential to obtain large areas in order to win. Small anchors are easy to make, but they control very little territory. In the next diagram, the black anchor used nine pieces to control 16 hexagons. The white anchor used just three pieces but controls only one hexagon.

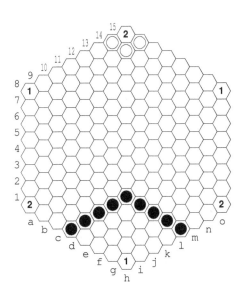

The next example shows a final position with dead stones marked with "a". White possesses 41 hexagons and one prisoner (adding up to 42 points). Black has surrounded 30 hexagons and has three prisoners (summing to 33 points). White wins the game.

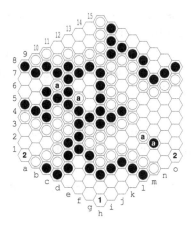

Anchor may suffer from a mirror strategy where the first player drops a stone in the middle of the board and then imitates the second player's last move. This problem can be solved by using a slightly asymmetric board, such as this one below.

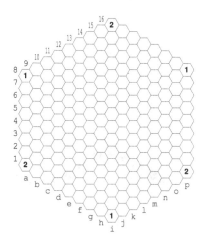

Reference

http://home.fuse.net/swmeyers/anchor.htm

Annuvin

This game, invented by Jeff Roy in 2001, has an interesting feature: when a player loses stones, all the remaining become more powerful. This produces a tension between capturing enemy stones and achieving a good position to one's own stones.

Materials

A hexagonal board with four hexagons per side and six white and six black stones. Initial setup:

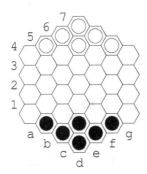

Rules

On each turn, each player moves one friendly stone. A stone may move along a chain of adjacent hexagons with a maximum of N cells.

The number N depends of how many friendly stones are still on the board.

$N = 1$ with six stones, $N = 2$ with five stones, $N = 3$ with four stones, $N = 4$ with three stones, $N = 5$ with two stones, and $N = 6$ with one stone.

If piece A moves to a hexagon occupied by an enemy piece B, piece B is captured and piece A may continue moving if it does not achieve its maximum.

Goal

A player that captures all enemy stones wins the game. A player that still has six friendly stones on the board also wins if he captures all but one enemy stone.

Notes

The second way to win prevents a tactic where one player sacrifices all but one piece in order to get a very powerful piece (able to move six hexagons) against an army of weaker pieces.

When captures start, players gain moving range. It is necessary to keep the army spread across the board to prevent multiple captures in a single turn.

In the next example, it is White's turn. He moves e6 (which can move up to three hexagons since there are still four white stones) and captures e4 and f4. Black replies by moving b3 (which now can move four hexagons) capturing d6 and d7. White captures c1 and d1 with f4. After these moves, Black has a desperate position. He has just one stone (which can move up to six hexagons) but the white army is widely spread, so Black cannot capture them all. White wins the game.

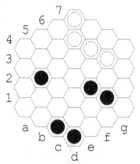

Annuvin lacks in strategy, but it has a strong tactical component, which is found in mobility and position. In the next example Black has the next move and a winning position based on his two groups. How can he bring this game to a close? Piece f5 captures d6. Next, one of the remaining white stones can move five hexagons, which is not enough to capture all black stones. If White uses b4 to capture d1 and c2, Black replies by capturing one of the white stones (and winning in the next turn). If he captures f4, Black makes the same type of move.

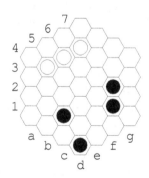

Campaign

This game was invented by Chris Huntoon in 2001. The rules are inspired by Gomoku (see p. 62) and by the ancient medieval puzzle Knight's Tour. The goal there is to move the knight over all the squares of the board, passing on each square once and only once. Here, the game represents a race between two pieces to make a five-in-a-row.

Materials

A square board with ten rows and ten columns, one white and one black special pieces (e.g., chess knights), and approximately 40 white and 40 black pieces.

Rules

Initially, one player drops both knights on empty squares. The other player chooses his color and starts.

On each turn, each player moves his knight to an empty square and drops a friendly stone on the square where the knight started his move. A knight moves like a chess knight, i.e., to one of the nearest empty squares not in the same row, column, or diagonal. The next diagram shows possible moves (the marked squares) of each knight.

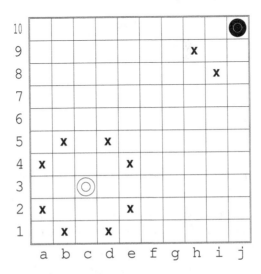

Goal

The player that first achieves a five-in-a-row with friendly pieces wins the game. A player may also win by stalemating the opponent's knight.

Notes

A player may win by making a line with more than five pieces.

Do not forget that there are two ways of winning Campaign. The first (making a line of five) is easier and should be the main focus of the player's strategy. However, if the adversary positions his knight in a crowded corner, the knight's mobility decreases, and it may be easy to stalemate it. An example of this situation is presented in the next diagram:

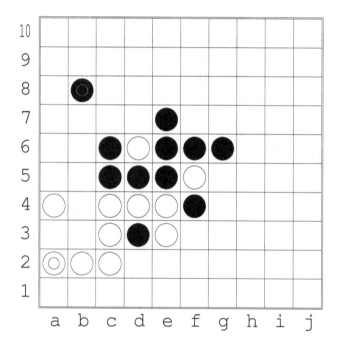

White moves to b4, threatening to win in the next move (with a horizontal five-in-a-row from a4 to e4. But Black wins the game by moving to the only accessible square to White, a6. White cannot move and loses.

This example shows that mobility is very important to prevent stalemates, but also to keep several paths for line making. This aspect is even more relevant because a player can only move a single piece: the knight.

Preventing the opponent's knight from moving to certain board sectors can be decisive.

As in Gomoku, a player must try to build open lines of four (i.e., lines of four friendly pieces where the extremities are adjacent to empty squares), in order to ensure multiple winning options. Since the knight's mobility is limited, it is hard for a player to prevent a double threat like this. Observe the next diagram, where the numbers represent the moves order:

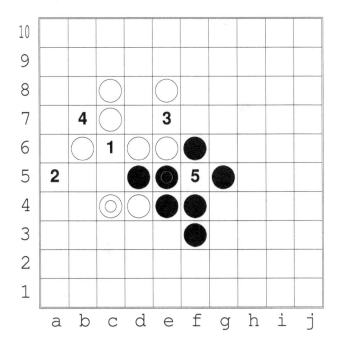

Black moves to c6 making an open four (with squares c6, d5, e4, f3) threatening a win at b7 or g2. White may prevent b7 but he is powerless to prevent Black's next moves (e7 and then f5).

Dispatch

This game was invented by Chris Dissemble and João Neto in 2002. The idea is to distribute groups of up to five pieces across the board in order to isolate and control as much territory as possible.

Materials

A square board with 13 rows and 13 columns, and approximately 75 white and black pieces.

Initial setup:

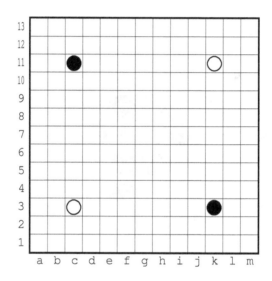

Definitions

Group — A set of friendly pieces orthogonally connected.

Liberty — A group has liberty if at least one of its pieces is adjacent to an empty square.

Rules

On each turn, a player drops a group (between two and five pieces) on a set of empty squares, as long as at least one of those squares is in a line with a square already occupied by a friendly piece, and both squares are separated by a line of zero or more empty squares. After the move, every group without any liberties is removed from the board.

44

A player may pass his turn.

Pie rule — The first player, on the first turn, may only drop a group of two or three pieces.

Goal

When both players pass consecutively, the game ends. Each player adds the number of friendly pieces (one point for each) and the number of squares on which his adversary cannot drop groups on (one point for each square). The player with the larger sum wins the game.

Notes

The next diagram presents the first two turns of a Dispatch game. Black moved and dispatched a group of three stones (they are limited in this first move by the Pie rule) trying to isolate the white stone at k11. White, instead of defending k11, attacks the black stone at k3. Black defends k3. White's 4th move tries to give maneuvering space to his threatened stone and, at the same time, control of row 13.

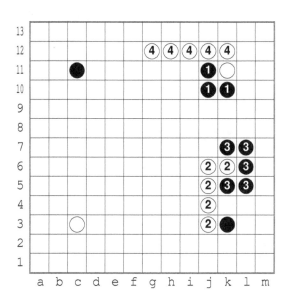

This is a territory game where concepts such as area and liberty (which lead to capture of enemy groups) are similar to Go (see p. 51). It is important to maintain the largest possible number of empty squares around friendly groups to prevent enemy attacks.

The capacity of dispatching groups is limited by the player's board position. So, every capture implies a decrease of that capacity, increasing the potential enemy territory.

Since part of the final sum is the number of friendly stones, it is advisable to dispatch groups of five stones (which results in five points). However, there are positions where a dispatch of five would remove too many liberties. In the next diagram, Black dispatched four stones (the marked stones). If he had dispatched five stones (occupying m13) his group would have lost its liberty.

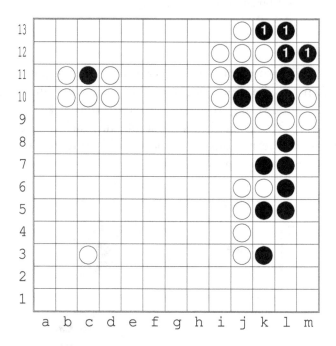

Epaminondas

This game was invented by Robert Abbot in 1963. Its original name was Crossings and was first published in 1969 in [SAC]. Crossings was played on an 8×8 square board. Its rules were revised and published in 1975 under the new name Epaminondas, a Theban general, inventor of the phalanx used to defeat Sparta in 371 B.C.

Materials

A square board with 12 rows and 14 columns, 28 white and 28 black pieces.

Initial setup:

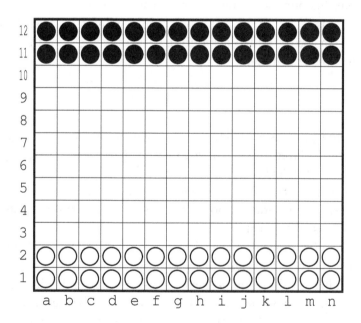

Definitions

Phalanx — an orthogonal or diagonal line of one or more adjacent friendly pieces.

Rules

On each turn, each player moves one friendly phalanx.

A phalanx moves over the line that defines it with a moving range up to the number of its pieces. For example, a phalanx with four pieces can move up to four squares along its line.

If the phalanx, while moving, encounters an enemy phalanx with a smaller number of pieces, that phalanx is captured. In this case, the friendly phalanx stops with its first piece on the square of the first enemy piece of the captured phalanx. In all the other cases, the moving phalanx must always move over empty squares.

Capturing is not mandatory.

Goal

If a player, at the beginning of his turn, has more friendly pieces in his last row than his opponent does, he wins the game.

For example, if White, at the beginning of his turn, has three white stones in the last row, and Black only has two black stones in his last row, then White wins the game.

Notes

An isolated piece is a phalanx of size one. It can move to any adjacent empty square. Isolated pieces cannot capture (because captured phalanxes must always be smaller).

It is not mandatory to move an entire phalanx. A player with a phalanx of five stones, may decide to move, say, just the first three stones (i.e., he moves a phalanx of size three).

Phalanxes may move to either direction along the line (e.g., a horizontal phalanx may move to the right or to left).

The winning condition is verified before the player begins his turn. If, after the move, a player has more pieces than the adversary, his adversary still has one chance to balance the position and continue the game (either by capturing some enemy pieces from his first row or by adding friendly pieces to his last row).

The next diagram shows some phalanx moves. Starting the game, White moved one square the phalanx defined by f1-g2, putting the first phalanx stone at h3. We describe this move as f1, g2-h3. Next, Black moved e12, f11-h9. White replied with a vertical phalanx, from h1 to h3, moving it to the maximum of three squares, i.e., h1, h3-h6. The result of these moves was:

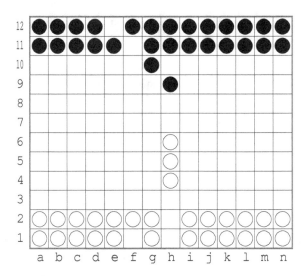

In the next diagram there is an example of a phalanx capture. The black phalanx h12, h9, with four stones, moves three squares down until it finds the first white stone of a phalanx with size three. Since the white phalanx is smaller, it is captured and removed from the board. If the white phalanx had four or more stones, this move would be invalid.

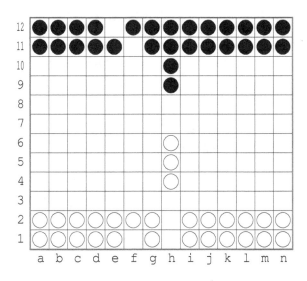

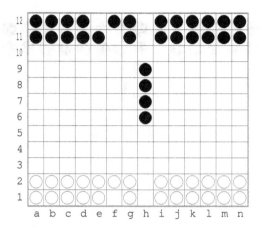

The next illustration shows a position where White has the advantage. The horizontal white phalanx in the first row is made of seven pieces, it is able to resist the attacks of the two potential black phalanxes (e5, c3 and e5, e3). But the horizontal black phalanx at row 12, with six pieces, cannot avoid the white double attack (a8, a9 and j5, j8), because these are separated by eight columns. White starts with j5, j8-j12, capturing one black stone. Black must capture that white stone at row 12 (or else, he would lose immediately) and White continues these captures, eroding the black phalanx. When these moves end, the other white phalanx at a8, a9 can move to row 12 without resistance.

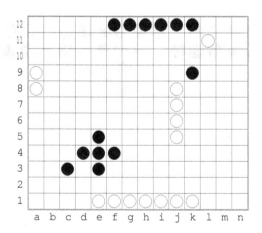

50

Go

Go is a traditional Oriental game. It appeared in China more than 2,500 years ago, was introduced in Japan around 800 AD, and is very popular in both countries. It is a game of influence, with simple rules, but with a remarkable strategic depth. In Chinese antiquity, it was one of the four arts taught to the noblemen. (The others were music, calligraphy and painting.) The game dynamics simulate a war with several localized battles which spread to the entire board. The accumulated knowledge through the centuries, in literature concerning opening theory, tactical and strategic ideas, is similar to chess.

Materials

A square board with 19 rows and 19 columns, approximately 150 white and 150 black stones. It is also common to play on boards with size 9×9 or 13×13 for faster and less strategic games.

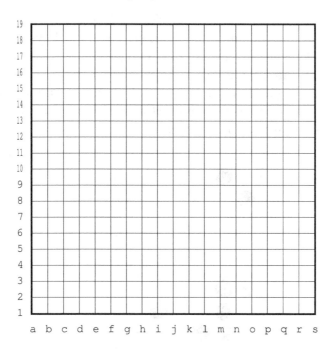

Definitions

Group — one or more friendly stones orthogonally connected.

Group liberty — a liberty for a group is an empty intersection orthogonally adjacent to at least one of the group stones.

The next diagram shows a group of seven black stones (with seven liberties) and a group of six white stones (with one liberty).

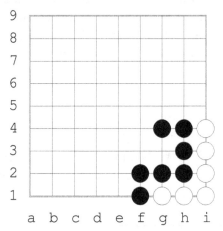

Territory — a set of empty intersections surrounded by a chain of stones of only one color and, maybe, by board edges. Both orthogonal and diagonal connections are permissible in determining this surrounding chain.

The next example shows three territories: One black territory with one intersection (at a9); another black territory with nine intersections; and one white territory with four intersections.

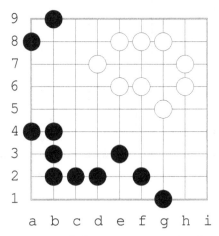

The number of stones needed to create territories is smaller at corners and larger in the middle of the board, as we can see in the previous diagram.

Rules

By tradition, Black starts. On each turn, each player passes his turn or drops a friendly stone in an empty intersection.

After the drop, all enemy groups without liberties are captured and removed from play. The captured stones are called prisoners.

A stone cannot make a suicide, i.e., it cannot be dropped in a group that, after the drop, loses all its liberties unless some captures are made.

Ko rule — A player cannot repeat the board position of the previous turn.

Goal

When both players pass consecutively, the game ends. Then all stones inside enemy territory that cannot survive and eventually would be captured are removed from the board. Each player adds the number of intersection he controls (territory) with the number of his pieces on the board. Whoever has the largest total wins. (If these numbers are equal, White wins.)

Notes

The number of liberties gives a measure of a group's strength. A group with just one liberty can be immediately captured. Players should try to avoid that and increase the number of liberties of friendly groups while trying to gain influence over the board. The next diagram shows a white group with four liberties. If White drops a stone at c3, the group will have three liberties. If White drops a stone at f4, the group will have five liberties. So, according to this criterion, f4 is better than c3.

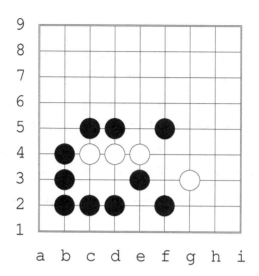

The suicide rule implies that a stone can be dropped at an intersection without liberties providing that it produces some captures. This happens because, after those captures, the stone will be adjacent to empty intersections. In the next example, it is valid to drop a stone at i5 because the group of six white stones is captured.

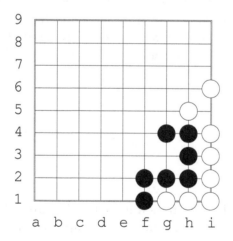

The suicide rule allows the essential pattern in Go, the living groups. A group is alive if the adversary cannot capture it at any circumstance.

The next diagram shows some living groups. In all of them, there are at least two separate territories. To place a stone in one of these territories is invalid. (It would be suicide.) The only way to capture them would be to drop two stones at the same time over those territories, which is impossible according to the rules. These territories are usually called eyes. So, a group is alive if it has two eyes.

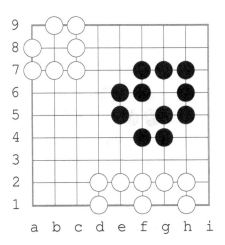

The next diagram shows groups that are not alive. The black group at the upper right corner just has one liberty at i9. If White drops a stone there, the group is captured. The bottom white group has two liberties. However, if Black drops at f1, the white group is in danger. Even if White drops at g1 (capturing f1), the group will only have one liberty. Then Black can capture the entire group by playing again at f1.

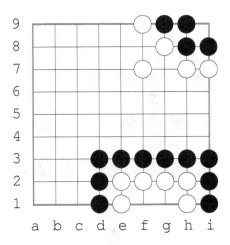

The Ko rule is essential to prevent positions where both players would endlessly repeat the same sequence of moves. In the next example, Black played at e4 capturing a white stone at e5. By the Ko rule, White cannot capture e4 by playing e5 because it would result in the previous board position. White must play elsewhere.

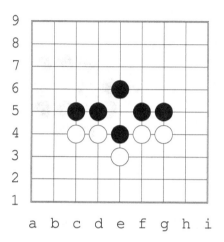

Notice that, on the next turn, White can play e4 unless Black has played there in the meantime.

The next diagram shows the position from a finished game. The two players just passed.

All stones not belonging to dead groups inside enemy territory are removed. In this example, there are three white stones (i2, b7, b8) and two black stones (f8, h6).

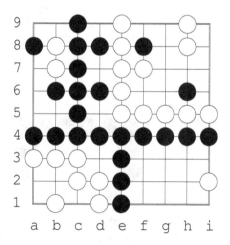

Next, players count their territories. Each player has two. White's left territory has four intersections and the right one has 13. Black's left territory has eight while the right has 12. There are still neutral areas not belonging to either player (like d3 or d9) which are not counted.

So, White has 19 pieces + 17 (intersections) = 36 points.

Black has 20 pieces + 20 (intersections) = 40 points.

Black wins the game.

The first player (Black) has a considerable advantage, so it is common that the second player should start already with some points. (This value is called Komi.) A typical Komi is 5.5. (The decimal part is to avoid ties.)

Gogol

The inventor of the game is unknown.

Materials

A square board with eight rows and eight columns, eight black stones (soldiers) and one black King, eight white stones (soldiers) and one white King.

Initial setup:

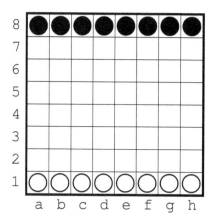

Rules

Initially, each player drops his King in an empty square, provided that the resulting position is valid. (See below.)

On each subsequent turn, each player moves a friendly piece, provided he generates a valid position.

Soldiers may move to any empty square.

Kings move along an orthogonal or diagonal line of empty squares (like the chess queen.)

The soldier and the King may capture adjacent enemy pieces by jumping over them and landing on the immediate next square (which must be empty.) Capturing is not mandatory, and it is not allowed to capture more than one piece in a single turn.

Invalid positions — there are special positions not allowed by the game rules:

1. The King must not be on the 1st or the 8th column while adjacent to a friendly soldier on that same column.

2. A King must not be in his 1st row while adjacent to a friendly soldier on that same row.

3. A King must not be in his second row while adjacent to a friendly soldier on the first row.

This diagram shows some invalid positions:

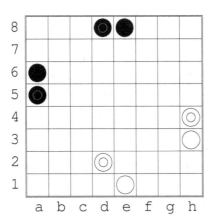

Goal

The player that moves his King to the last row or captures the enemy King wins the game.

Notes

Corners are very strong positions. A piece in a corner cannot be captured.

Players should be very careful not to allow many open spaces in their first row. The enemy King may take the chance to put itself in a position where it makes a double threat, i.e., he is able to move to two different squares in his last row.

The next diagram shows a black King that threatens to move to g1 and, at the same time, attack the enemy King which must defend itself (while opening a path to b1). White has lost this game... .

59

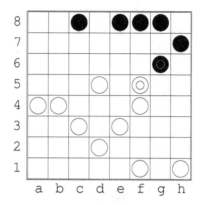

When a player loses some soldiers, he must prevent the enemy King from going to the third row, if he has no pieces there. If that happens, the adversary can force, at least, a draw by simply moving to squares without a soldier in the first row of that column (the enemy will keep a constant winning threat that the player cannot break).

The next diagram shows this case. If the black King moves to d3, White must defend by moving a pawn to d1 (moving a pawn to d2 would allow the King to capture it with d3:d1, winning the game). Then the black King would move to g3 and so on... .

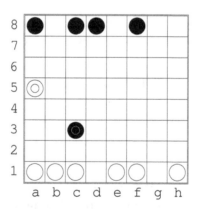

In the next diagram, White tries a sacrifice by moving his g1 soldier to h6. Black faces a dilemma: if he captures it with his King with h5:h7, White replies h3-h5 threatening to put a soldier at h8 (Black must move his King to g6 and White wins with h5:f7). So Black cannot accept the sacrifice and must move to another square. However, White may use his h6 soldier to attack the black King while perhaps opening the h column to his own King.

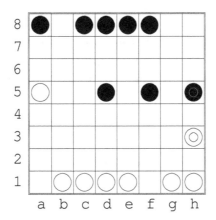

Gomoku

Gomoku is a traditional Japanese game played on a Go board (see p. 51). Its origins, like Go's, are Chinese and derive from the ancestral game *wuzi*. Its complete name is Go-moku Narabe, literally meaning "five stones in a line."

Materials

As in Go, a square board with 19 rows and 19 columns, and approximately 75 white and 75 black stones will be needed.

Rules

On each turn, each player drops a friendly stone on an empty intersection. Black starts dropping a stone on the central intersection.

Goal

The first player with an orthogonal or diagonal line of exactly five consecutive friendly stones wins the game.

The next diagram shows a diagonal line with five black stones.

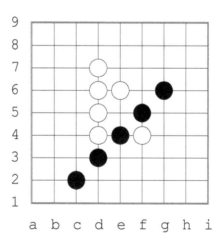

Notes

The first player has a great initial advantage. There is a more refined variant, called Renju, which uses a complex set of procedures to achieve fairness. Herein, we propose a simple Pie rule to balance this advantage:

One player drops three black stones and two white stones on the board. The other player chooses colors. White starts.

There are patterns that result in certain victory. If a player gets a line of four stones where both extremities are adjacent to empty cells, the adversary cannot stop in just one move this double threat. This pattern is called an open four. The next diagram shows one white open four:

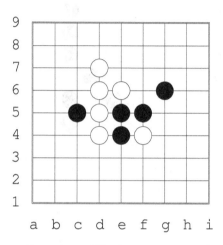

White already won the game. Black cannot stop both winning moves at d3 and d8.

In the next example, Black played at g6 and, in the next turn, wins the game either at g7 or i8. This type of pattern is called a double closed four.

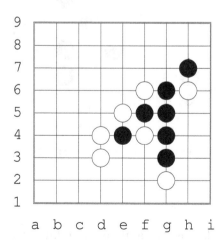

It is implicit in the goal that lines of six or more stones are not valid to win (though western players almost always ignore this restriction).

In the next diagram, White wins with c3. This is the combination of a closed four and an open three. Black must avoid the immediate threat at b3. This gives White enough time to drop at b2, winning due to an open four.

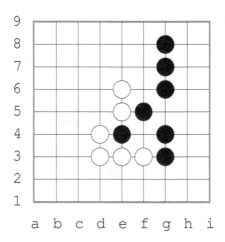

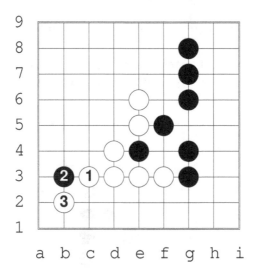

Variants

There are a lot of different variants of Gomoku. (Tic-tac-toe can be seen as one.) We present two variants which are, at least, as interesting as the original game.

Gomoku Ninuki

The rules are like those of Gomoku, but here we have a custodian capture rule for pairs of enemy stones. The player that makes a five-in-a-row or captures ten stones wins the game.

In the next diagram, White can capture four black stones if he plays at d5 (there are two custodian captures at the same time).

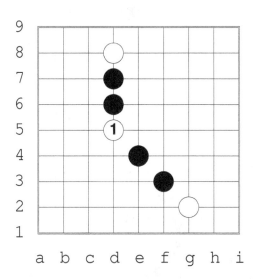

Gomoku with Dagger

To describe this variant, we need a definition.

Dagger — the capacity of playing twice in the same turn. It is not valid to use the dagger to immediately win.

The rules are like Gomoku's, but the second player starts with a dagger. When a player uses the dagger, he must give it to the adversary. A player cannot use the dagger immediately after the opponent has used it except to protect himself against an immediate loss, i.e., on the opponent's next move.

It is White's turn in the next example. White has the dagger. He plays the dagger (passing it to Black), dropping two stones at f4 and f5. Black must defend by playing the dagger again with h3 and c8. So, White receives the dagger again (but he cannot use it on this turn, since he used it on the previous one). Yet, it is now enough to play at f6, winning because of an open four.

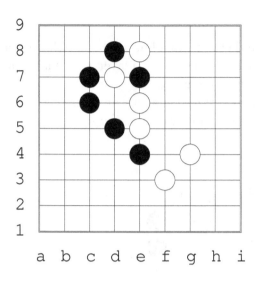

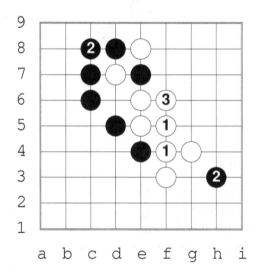

Notice that, initially, White could not win simply by dropping f5 and h3 because it is invalid to use the dagger as the winning move.

Gonnect

This game was invented by João Neto in 2000. The idea is to mix the strategy of Go with the tactical richness of Hex, resulting in a connection game where the notion of area is essential.

Materials

A square board with 13 rows and 13 columns with approximately 75 white and 75 black stones.

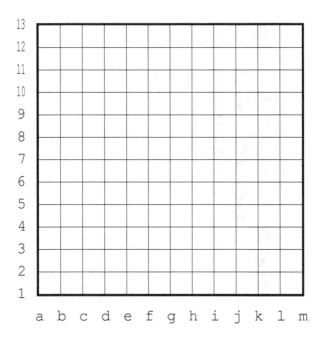

Rules

The rules of Go (see p. 51) apply except:

A player cannot pass his turn.

There is a Pie rule: the first player drops a friendly stone and the second player may swap colors.

Goal

A player wins if he manages to have a chain of orthogonally adjacent stones connecting two opposite board edges (either from left to right or top to bottom). A player may also win if he stalemates the adversary.

Notes

The first Go rule exception (players may not pass) implies that all structures, including living groups, may be captured in the long run. This happens because players must always drop a friendly stone at their turn. Since players cannot commit suicide, they eventually run out of good empty intersections and must start to occupy their own area.

A suicide is always illegal even if that drop would create a winning structure. The next diagram shows an example of this type. White cannot play at a1 because the group would have no liberties.

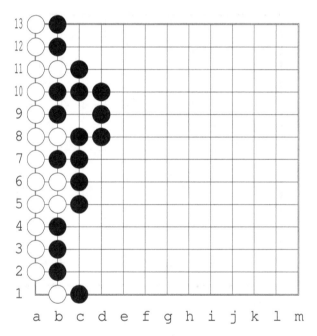

In the opening moves, each player tries to establish a set of friendly pieces that provides good connection possibilities between edges. Usually, the stronger this connection becomes, the slower it is to make it. Here are some types of connections (shown in the next diagram, from left to right, and up to down):

- strong and very slow

- equally strong and twice as fast

- amost as strong as the first two

- fast but not that strong

- slow (but not very slow) and strong

- fast and strong

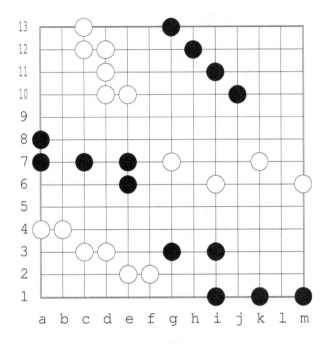

Players may mix connection types when creating the backbones of their groups (e.g., the 5th connection is a mix of the first two). Everything depends on the neighboring stones. If a connection is threatened by nearby enemy stones, a stronger, slower connection may be needed. On the other hand, if there are friendly stones nearby, connections may be made faster.

Sometimes there are connection races in Gonnect. This means that both players are almost winning and the number of stones to achieve victory decides who wins. If a player is behind in the race, he may drop stones near the adversary's structure to delay its progress.

In the next example, Black is losing the race. If he drops a stone at b4, he will delay White, who will need two moves (b3 and a4) to capture b4 and keep its structure in the race.

69

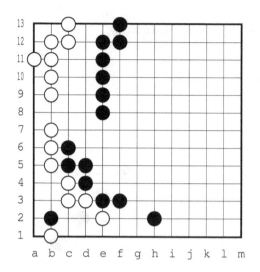

Typically, there will appear blocking positions, i.e., patterns where no player will be able to achieve a winning connection directly. From that moment, it is crucial to dominate as much area as possible. The reason is to prevent the adversary from having many good dropping options and make him be the first to start occupying his own area (and thus eventually to destroy his living groups). When a living group is captured, the owner will lose the game almost surely. Usually, when both areas are consolidated, it is easy to see who will win the game.

In the next diagram, a blocking happened between a11 and b10. Yet, White's influence is so great that he probably has already won the game.

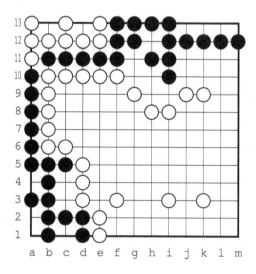

70

Gonnect is not merely a Go variant (though obviously it was inspired by it). Both games have different goals which result in different strategies. The main difference occurs in the opening moves. There is a much easier task to achieve (connect sides) which provides clearer goals and quicker, meaningful tension, especially to beginners.

A sample game: c3 f6, f3 g3, g4 h3, h4 j3, j4 l3, f4 l5, h6 j9, k7, l7, k8 m8, l9 m10, m9 n9, l10 m11, l11 m12, l12 m13, l13 l8, k9 j6, j7 (White resigns) 1 - 0.

Final position:

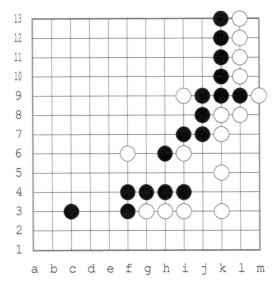

Black won the connection race. He needs four stones to win while White needs six. There is no way to delay Black's structure.

Havannah

This game was invented by Christian Freeling in the 1980s. It is a game with simple rules and multiple goals but strategically deep.

Materials

A hexagonal board with ten hexagons per side, 100 white and 100 black stones.

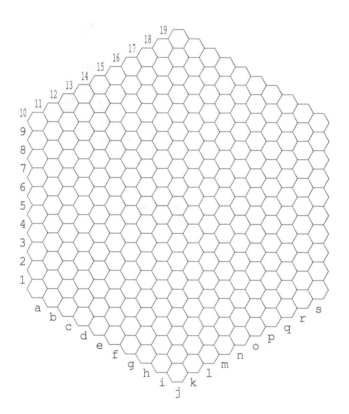

Rules

On each turn, each player drops a friendly stone on an empty hexagon.

Goal

The player that first achieves one of the following patterns wins the game:

- a Ring — a chain of friendly stones that surrounds at least one hexagon (which may be empty or occupied by either player).

- a Bridge — a chain of friendly stones that connects two corners.

- a Y — a chain of friendly pieces that connects any three edges. (Corners do not count as board edges.)

The next diagram shows one black ring (surrounding three hexagons), a white bridge (connecting corners a10 and j19) and a black Y (connecting the east, southeast and northeast edges).

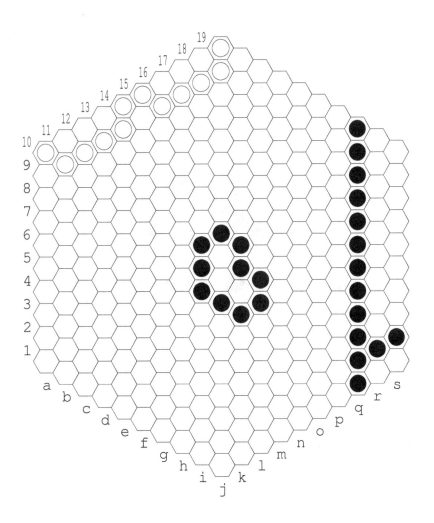

Notes

This is a game of connection, but victory lies in the influence that the player's stones have over the board. The center is especially important; it facilitates the access to all edges and provides paths to build bridges and Ys.

To win with a bridge, the player must drop two pieces in corners, two hexagons very far from the center. Also, it is easy for the opponent to drop a piece in the corner that the player did not occupy. Bridges can be seen more as potential threats than valid goals to the player's strategy.

Players try to create safe structures, i.e., chains that lead to victory no matter the opponent's response. This means to see who is the first to conclude his safe structure. The next diagram shows a safe structure that guarantees a ring:

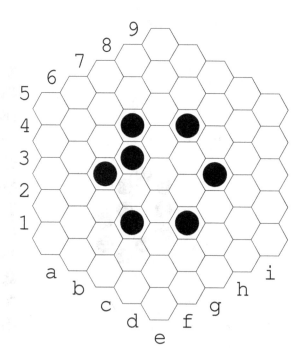

There are several tactics to create and to block this type of structure. One important point is to know how to extend a structure in a specific direction. The easier way is the simple extension. (Check the upper sector of next diagram.) If Black plays at d7, the connection with b6 is guaranteed. If White plays c7, he replies c6, and vice-versa.

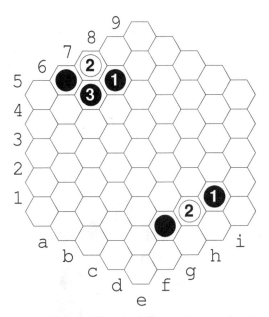

The other extension (from f3 to h5) is easily cut (with g4).

Other fast extensions can be cut. In the next diagram, Black tries to extend f3 three times and forces White to cut consecutively. However, after c3, Black has now a very strong structure. An extension is more important as a threat than as an effective connection to the extended stone.

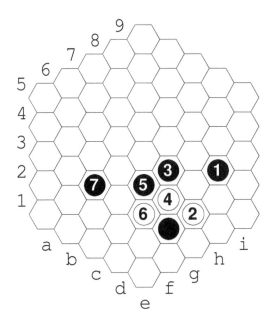

There are structures that seem safe but are not. The next position shows one. White is able to destroy Black's threat. It seems that, until move 9, White is helping Black to make a ring, but, after move 11, the ring is destroyed. Notice that move 10 could not be at c3, because then White would play d5 and win the game with a white ring.

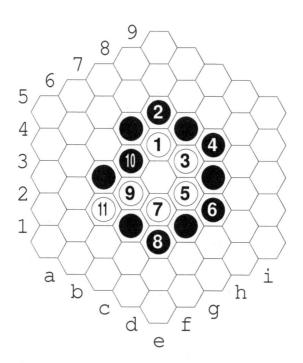

Reference

http://www.mindsports.net/Arena/Havannah/

Hex

The game of Hex was invented at least twice: once, by the Danish poet and mathematician Piet Hein in 1942 and the other by the American mathematician John Nash in 1948. However, it was Martin Gardner who popularized it in the pages of *Scientific American.*

Materials

A diamond-shaped board with hexagonal cells, with 11 cells on each side, 61 white pieces, and 61 black pieces.

Definition

Adjacency — Two pieces are adjacent if the hexagons they occupy share an edge.

Group — A set of same-colored pieces connected by orthogonal adjacencies.

Rules

Hex is played in a board like the one in the illustration. There are two players, White and Black.

Each move consists of dropping a piece in an empty hexagon.

The Pie rule holds: on his first move, the second player can change colors (using the move his adversary just did).

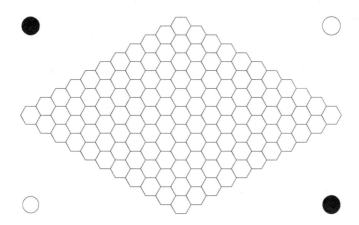

Goal

Black wins if he can form a group of black stones connecting the southeast and northwest edges.

White wins if he can form a group of white stones connecting the southwest and northeast edges.

Notes

The Pie rule is very important in Hex. The Berkeley mathematician David Gale showed that this game cannot end in a tie, and John Nash proved that, without this special rule, the first player must have a winning strategy. However, for nontrivial dimensions of the board (including 11×11), nobody knows that strategy. His proof is an existential one, and it uses an argument of reduction to contradiction. We recall it here.

We mentioned already that the game cannot end in a tie. Therefore, either the first or the second player has a winning strategy.

Let us assume, for one moment, that the second player, playing perfectly, can always win. Under this condition, the first player should play his first move at random and act as if he were the second player. As the second player has a winning strategy, he can use it now. If, at some moment, he needs to play where he already did, he just plays at random again. This strategy must give him the victory. In short: assuming that the second player has a winning strategy, we conclude that the first player has a winning strategy! This contradiction shows that we cannot assume that the second player has a winning strategy, thus, it is the first player who wins.

This line of reasoning is now classic and is widely known as a *stealing strategy argument*.

As this is a connection game, it is important to know how to extend a group. A move to a neighbor cell seems too slow.

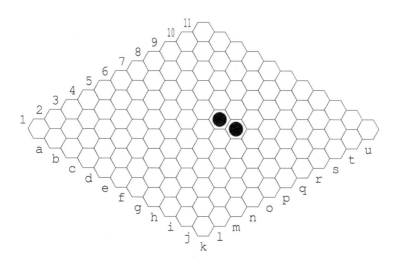

Here the pieces in l7 and m7 are too close to each other, hardly helping to connect the edges.

On the other hand, a big jump can be cut by the adversary:

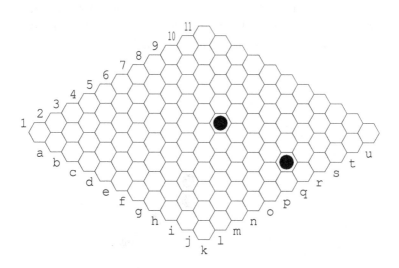

In this case, the pieces l7 and p7 can be effectively separated if White plays in n7.

An example of a good connection, not too far-reaching, is a *bridge*, which consists on having two stones that share two neighbor cells:

79

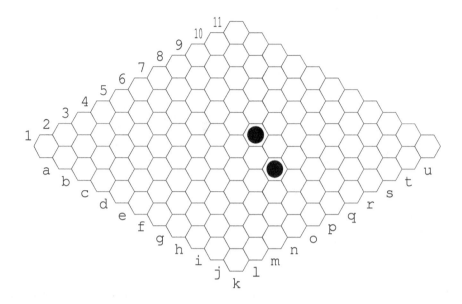

Here, if White tries to cut, playing l6, Black replies in m7, and if White plays in m7, Black replies in l6. The two stones are as good as if they were already connected.

Defensive moves must also be well thought out. For instance, trying to block an adversary group should not lead us to play too close to the adversary group, because then it can extend easily:

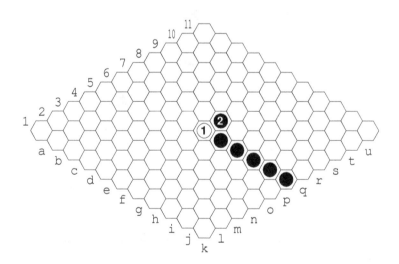

Trying to block in k6 is met with a move in l7. Even at a bridge distance, a block is not effective:

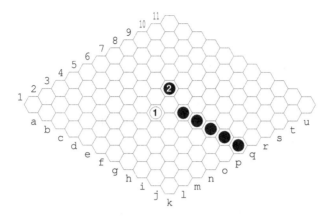

If White takes j5, Black replies in k7. Good moves are to be found at a distance:

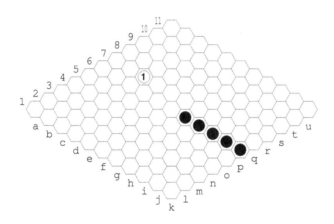

References

Browne, C., *Hex Strategy: Making the Right Connections*, A. K. Peters, 2000.

Gale, David, "The Game of Hex and the Brouwer Fixed-Point Theorem," *American Mathematical Monthly* 86(10):818–827.

Hexy, in http://home.earthlink.net/~vanshel/

Hobbes

This game was invented by Dan Troyka at 2002. It is a game where two kings with powerful moving ranges maneuver neutral pieces in order to stalemate or capture the adversary's king. The name and dynamics came from the English philosopher Thomas Hobbes. His work *Leviathan* tried to show that the people should gravitate around absolute monarchy to avoid falling into the violence and conflicts of a supposedly natural state.

Materials

A square board with five rows and five columns, one white and one black stone, and ten neutral stones.

Initial setup:

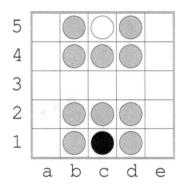

Rules

On each turn, each player executes the following order of moves:

1. Optionally moves his king to an orthogonal adjacent empty square. This movement may be repeated as many times as the player likes. If the king becomes orthogonally adjacent to the opponent king, he may capture it.

2. The player must do one of two actions:

a) Orthogonally push one adjacent neutral stone by one or more empty squares.

The next diagram shows the white king pushing d3 by one square:

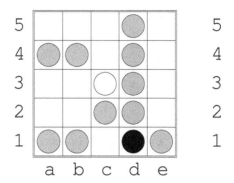

b) Orthogonally pull one adjacent neutral stone by one or more empty squares.

The next diagram shows the king pulling d3 by two squares:

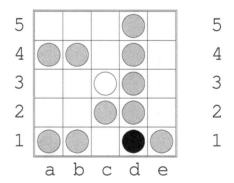

Goal

The player that stalemates or captures the opponent king wins the game.

Notes

There are two ways of winning: by capturing the enemy king or by preventing it from moving. The next diagram shows a capture example. The white king may move to d2 and push c2 to b2. (The white king will end the move at c2.) This way, the only possible black move is to push c1 to d1. On the next turn, White captures the adversary.

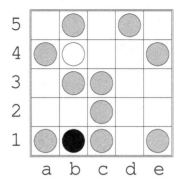

The next diagram shows a stalemate win. The white king moves to c3 and pushes the neutral stone at c2 to c1. Black cannot move and loses the game.

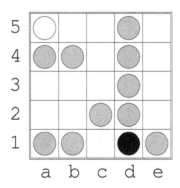

Each player should try to minimize the movements of the enemy king in order to reduce his tactical possibilities. On such a small board (it is possible to play Hobbes on larger boards) the first move that pushes the neutral stone in front of the king to the center cell is very strong and should be forbidden among stronger players, (alternatively, just include the pie rule in Hobbes).

It is White's turn in the next diagram. The winning move is to push c3 two squares to a3. This way, Black only has one possible move to push c5 to d5. White continues his attack by pushing b4 to b5. The black king can only push b5 or c4; either way he will be captured soon.

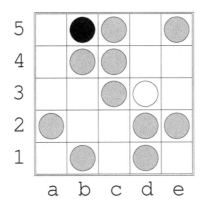

Intersections

The author of this game is unknown. This is a simple game where two special stones slide over their row/column in order to determine where the next stone is dropped.

Materials

A square board with six rows and six columns, 20 white and 20 black stones.

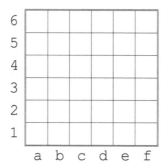

Definitions

Intersection — the square that intersects the row and column of the two special stones.

Rules

Initially, one player drops a friendly stone in the left column and the other drops a friendly stone at the upper row. (These are the special stones.) Neither can occupy the square a6 at any time. Then the second player drops a friendly stone at the intersection.

On each turn, each player slides his special stone (Black vertically, White horizontally) and places a friendly stone at the new intersection. Players must drop into an empty intersection. If a player tries to move and there are no empty squares, he may play to an intersection occupied by an enemy stone which is captured and replaced by a new friendly stone.

Goal

The player that achieves an orthogonal or a diagonal four-in-a-row wins the game.

Notes

This is a very tactical game; there is no time or space for a strategic plan to emerge. It is necessary to keep the player's lines open in order to have more options to move.

White should avoid almost full lines (and Black almost full columns). If this happens, the adversary will be able to capture friendly pieces since there are no empty squares on that line (column).

It is mandatory to move to empty intersections, so it is possible to force a certain move sequence that may result in a win. The next diagram shows an example. (The special pieces are marked with a circle). It is White's turn. He can only move to columns b or c. If White moves to c, he will drop a white stone at c2, and Black can win by moving to row 5 and dropping a stone at c5 because White is forced to move to column e placing a white stone at e5. Then Black makes a four-in-a-row by sliding to row 4 and placing a black stone at e4.

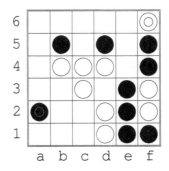

The next diagram shows a position of a game where White started by putting his special stone at c6 and Black at a3. Then, Black placed a black stone at intersection c3, White moved to column e, dropping a white stone at e3, and so on.

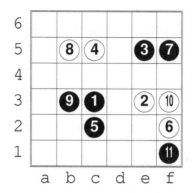

In this position, black move 11 is the only sensible one since a move to f4 would give White an immediate victory at d4. However, even with this move, White can win if he drops at c1. Black is forced to c4 (the only empty square at that column), which allows White to move to d4 and win with a diagonal four-in-a-row.

Iqishiqi

This game was invented by João Neto in 2003. The name is a derivation of IQI (ichi) which is number one in Chinese, OSHI which means push and QI (chi), game, that is, the game of pushing one stone.

Material

A hexagonal board with eight hexagons per side, one black and 75 white stones.

Initial setup:

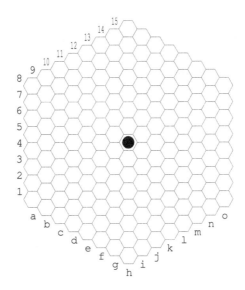

Definitions

Group — a connected set of white pieces.

Board edges — each player owns three non-adjacent edges. We use East to identify the player owning the east, northwest, and southwest edges. West is the player that owns the west, northeast, and southeast edges.

Rules

On each turn, each player drops a white stone on an empty hexagon from where it pushes the black stone (also called neutral stone).

The neutral stone is only pushed if the dropped white piece belongs to a group where at least one of its pieces (including the dropped piece) is in the line-of-sight from the neutral stone (i.e., there are no stones between the neutral and that group stone). If that happens, the neutral stone is pushed away from the group a number of hexagons exactly equal to the group size.

Goal

When the neutral stone is pushed to a board edge, the player owning that edge wins the game. If the neutral stone is pushed to a corner, the player that made the push wins. A player that stalemates his opponent wins.

Notes

The neutral piece cannot be pushed to the board edge if it cannot advance the exact number of hexagons equal to the pushing group size. The next diagram shows an invalid move 1. The neutral stone is in the line-of-sight (of stone X) but it cannot advance four hexagons (the group has now four stones) since the edge is only three hexagons away.

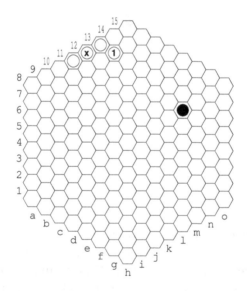

A move may be able to push the neutral stone in more than one direction (because the neutral stone may be in more than one line of sight from the group). The next diagram shows an example like that. Move 1 could push the neutral stone in any of three different directions to one of the marked hexagons.

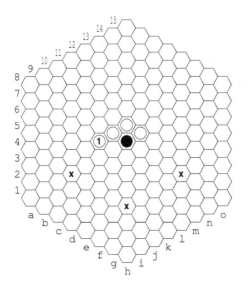

In the next diagram, it is East's turn. If he drops a stone at 1, there is a forced winning sequence. (Numbers show where the white stones are dropped and the letters show where the neutral stone is pushed.) Moves 2 and 4 are forced; there are no other options.

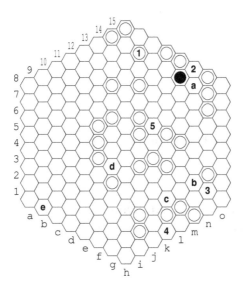

Jade

This game was invented by Mark Thompson in 2001. The game is based on Hex (see p. 77) where players may drop stones of either color to achieve their different and contradictory goals. Since stones are shared, the positional tension remains strong.

Materials

A hexagonal diamond board with nine hexagons on one side and nine on the other, 50 black and 50 white stones.

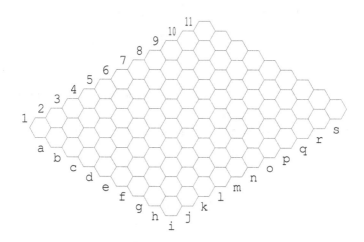

Definitions

Group — a connected set of stones of the same color.

Rules

Players are called "Cross" and "Parallel." Cross starts.

On each turn, each player drops a stone of either color on an empty hexagon.

Goal

Cross wins if there is a group (of either color) that connects all four edges.

Parallel wins if there is one white group and one black group connecting any two opposing edges.

Notes

Cross and Parallel have opposing goals, i.e., they cannot be satisfied at the same time. Like in Hex, there are no ties; one of the players must win.

The next two diagrams show two winning groups, the first for Cross and the next for Parallel.

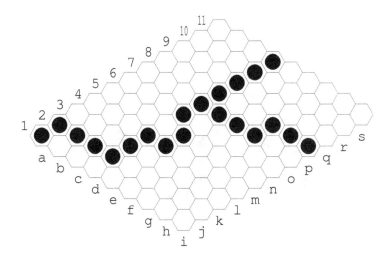

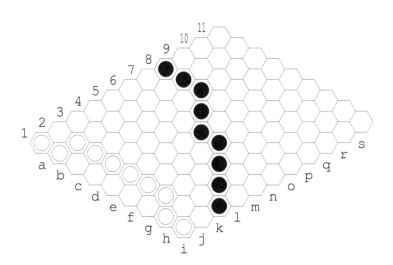

93

Players (especially Cross) must use both colors to keep as many open paths to victory as possible. It is usual that Cross should take the initiative, which Parallel tries to neutralize. The first version of Jade's rules used the same board as Hex, but Cross won an average of 75 percent of the recorded games. The two extra rows in the current board make it harder for Cross and balances the game.

There are positions in which it is possible to force the adversary to lose by playing in a crucial hexagon, no matter the color he chooses. The next diagram shows an example. Cross played a white stone at i8, forcing a decision at i7. According to the color played at i7, Cross plays a stone of the same color at k8, creating a very strong structure.

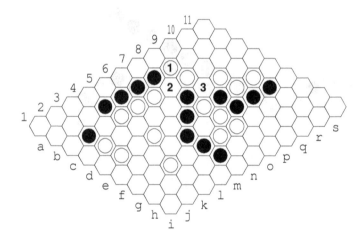

Lines of Action

This game was invented by Claude Soucie in 1969 and is described in [GG]. The game got the attention of the Artificial Intelligence community which produced software able to defeat the best players in the world. An example is Mona, a program from the University of Alberta, Canada.

Materials

A square board with eight rows and columns, 12 black and 12 white stones. Initial setup:

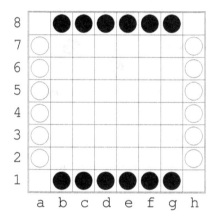

Definition

Group — an orthogonally or diagonally connected set of stones of the same color.

Rules

On each turn, each player moves one friendly stone.

A stone may move orthogonally or diagonally as many squares as there are stones (of either color) over that line (including itself). It is possible to jump over friendly stones. If a stone lands over an enemy stone, the latter is captured and removed from play.

Goal

The player that first connects all his stones still on the board wins the game.

Notes

An isolated piece is considered a group.

Initially, it is not easy to visualize all possible moves. There are only four lines to consider (horizontal, vertical and the two diagonals) for each friendly stone. Do not forget that your stones affect the enemy mobility. The next diagram shows all possible moves of stone d4 (stone b6 can be captured).

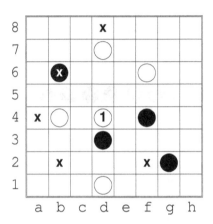

The next example shows a winning position for Black. All black stones are connected into a single group.

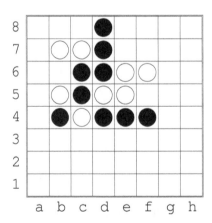

If a player's move creates a winning position for both colors, the victory is still his. In the next diagram, White moves f2 to c5 capturing a black stone. With this move, both players are reduced to a single group, but it is White's victory.

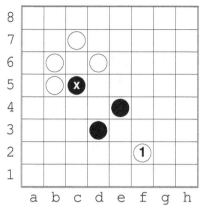

There is a tension associated to capturing enemy stones. An adversary with few stones has less mobility. However, it is easier for him to merge into a single group. In fact, a player reduced to one stone satisfies the winning conditions. This tension implies that sacrifices (where a player permits the capture of friendly stones) may be advantageous.

Players should choose a region to try to merge their stones. Hesitations that make stones go back and forth in an area give the initiative to the adversary.

Since stones cannot jump over enemy stones, it is possible to build walls that separate enemy groups. In the next diagram, the white group with stones a4 and a5 is separated by the black group that can unite its stones without major problems and so win the game.

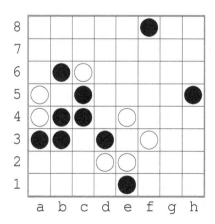

References

Abstract Games Magazine, 1.

http://www.andromeda.com/people/ddyer/loa/loa.html

Nex

This game was invented by João Neto in 2004. This is an example of an original game (in this case, Hex) modified by a mutator. Herein, the mutator is the use of a neutral stone drop or a swap rule.

Materials

A hexagonal diamond board with eleven hexagons per side, 50 white, 50 black, and 50 grey stones.

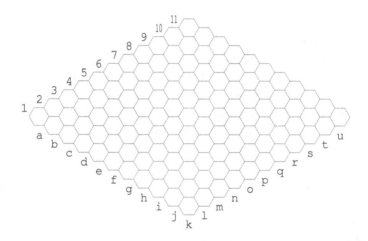

Definitions

Group — a connected group of pieces of the same color.

Rules

On each turn, each player should execute exactly one of the following options:

1. Drop a friendly stone on an empty hexagon, and drop a neutral stone on another empty hexagon.

2. Swap two neutral stones for friendly pieces, and swap another friendly piece for a neutral stone.

Goal

Black wins if he makes a group connecting the southeast and northwest edges.

White wins if he makes a group connecting the southwest and northeast edges.

Notes

Neutral stones are represented by grey stones.

Similar to Hex and Jade, both players' goals cannot be satisfied at the same time. There are no ties. So one player should always win.

The next diagrams show, respectively, a black win and a white win:

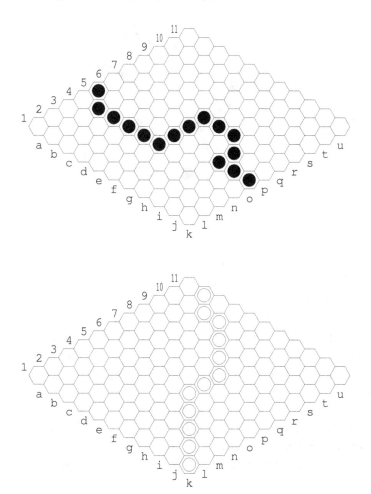

The use of neutral stones greatly increases the game complexity with a small increase in rule complexity. The tactical possibilities are larger since neutral stones are both a benefit and a problem shared by both players.

Whatever the options of White and Black, the number of black and white stones will always be equal at the end of each turn. The number of neutral stones, however, can change and be reduced even to zero.

The next diagram shows a real endgame, where Black resigned because he was unable to prevent the triple threat at k11, l11 and l10. Black can swap two stones, but there is always the third granting a White win.

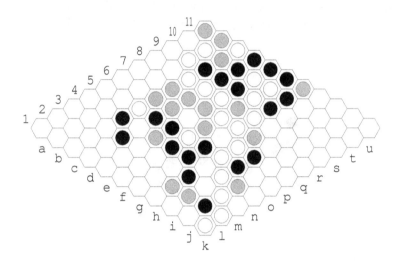

These games are full of tactical subtleties. An interesting feature is that there are no useless stones on board; it is always possible to recycle them to swap a pair of neutral stones.

When the number of neutral stones rises, the game reaches a critical mass that produces swap battles that may decide the winner. Forcing moves (i.e., making the adversary play at a specific point) are the easiest way of preventing the adversary from swapping important neutral stones on his next turn.

It is common the co-existence of two plans: one with friendly pieces and another based on neutral structures that mines the adversary groups and may provide an alternative path to victory.

In the extreme (and rare) case where the board is almost full, we apply the following conventions:

1. When there is only one empty hexagon and there are no neutral stones, the player just drops a friendly stone.

2. When there is only one neutral stone and no empty hexagons, the player converts the remaining neutral into a friendly stone.

100

3. When there is one neutral and one empty hexagon, the player drops a friendly stone on the empty hexagon and the opponent converts the neutral stone.

These conventions are required to completely define Nex and any other game where this mutator is applied. Actually, games like Gomoku, Y and Gonnect can be transformed into very nice variants with this concept.

Nosferatu

This game was invented by Chris Huntoon in 2001. The players possess different armies and different goals: Nosferatu and his vampires against a not so friendly set of peasants.

Materials

A hexagonal board with four hexagons per side, 12 white stones (the peasants), 17 black stones (the vampires), and one special black stone (the Nosferatu).

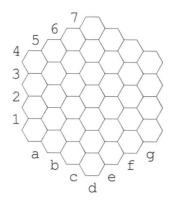

Rules

White starts with his 12 pieces. Black starts with 6, one of which is the Nosferatu. All the pieces start off the board.

On each turn, each player must do one of the following actions:

1. Drop a friendly piece on an empty hexagon.

   It is illegal to drop a piece on a cell where it could be immediately captured.

2. Move a friendly piece to an adjacent empty hexagon.

3. Capture an adjacent enemy piece by jumping over it and landing on the consecutive cell in that direction (which must be empty).

   The jumping piece may, optionally, continue to capture other enemy pieces by the same method in the same turn. It is not mandatory to maximize the number of captured pieces.

102

Capture has precedence over dropping, and dropping pieces has precedence over moving pieces.

The Nosferatu has some extra powers:

1. When moving, the Nosferatu may slide over a line of one or more empty hexagons.

2. When capturing, the Nosferatu may slide over a line of one or more empty hexagons before and after the jump.

3. Peasants captured by the Nosferatu may be removed from the board or transformed into vampires.

Goal

Black wins if he captures all the peasants. White wins if he captures the Nosferatu.

Notes

The precedence of capturing, dropping and moving imply that a player can only move when there are no captures and all his off board pieces have been already dropped.

The initial phase is fundamental and most times determine the final outcome. White should prevent an early conflict before all his stones are already on board. The number of peasants determines the moving range of the enemy. With lots of peasants, the vampires cannot move easily and are more prone to fall into traps. For White, sacrificing a peasant to capture a vampire is usually a good move.

Black should avoid placing the Nosferatu too soon. Because of its extra powers, the Nosferatu is a very powerful and yet a very vulnerable piece: powerful because it can capture many peasants in a sequence of long jumps, making them vampires; vulnerable since capturing is mandatory and it is not hard for White to force Nosferatu to capture one or more peasants and put itself in a losing position.

The next two diagrams show these types of situations. On the first, the vampires win with the black move e7:e5 (i.e., vampire e7 captures e6 and moves to e5). White replies with d5:f5 (d4:f6 is not better). Then, Nosferatu makes a sequence of ten moves: g4:g6, g6:d3, d3:d1, d1:d6, d6:b4, b4:e3, e3:e1, e1:a2, a2:c4 and finally c4:c2. Black won the game by capturing all peasants in a single move!

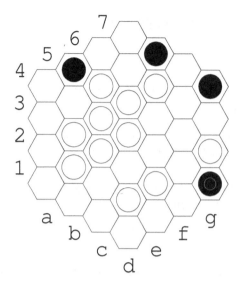

In the next diagram, White moves d3-c2. Black must capture with b2:d2. Next, White plays e2:c2 and the Nosferatu must capture a2:d2:g5 or a2:e2. In both cases, White can capture the Nosferatu, winning the game.

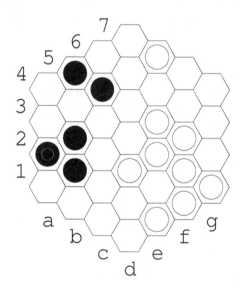

Pawnographic Chess

This game was invented by Bill Taylor in 2000. It is based on the rules of chess. The winner is the first player to promote a pawn.

Materials

A square board with eight rows and eight columns, eight white and eight black stones.

Initial setup:

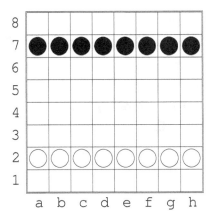

Rules

White starts. On each turn, each player moves one friendly pawn.

Pawns move like in chess.

They move to the forward vertical square if empty, unless if they are in the initial position where they can move one or two squares forward (always to an empty square).

They may capture an enemy pawn if it lies on a diagonal forward square. Captures are not mandatory.

As an example, the white pawn at e4 may capture f5 but cannot capture e5.

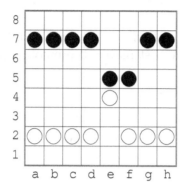

Pawns may also capture *en passant*, that is, a pawn which 'attacks' an empty square crossed by an enemy pawn that just advanced two squares, may capture that pawn, moving to that empty square. This move must be done immediately after the aforementioned enemy move.

The next diagram shows an example of an en passant capture:

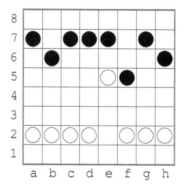

Black just moved f7 to f5. If White wishes, he may capture that black pawn en passant, moving the white pawn from e5 to f6.

Goal

The player that first moves a friendly pawn to his last row wins the game. A player may also win by stalemating the adversary.

Notes

Some notions from chess are useful, such as piece mobility or pawn structure. However, the domain of the center does not seem to be relevant. It is more important to possess en passant capture threats.

106

The next diagram illustrates these points. While White has the center, Black has two en passant threats (in columns a and g) that pin down the white pawns b2, f2, and h2. This position is desperate for White.

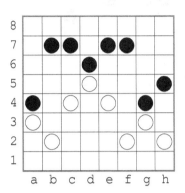

An uncapturable pawn is much more important in this game than in chess since it guarantees a win (unless the adversary also possesses a similar and better placed pawn). In the next example, both players have pawns like this; however, White wins the race even if it is Black's turn.

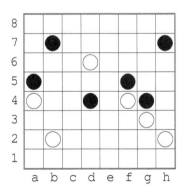

Many games end with a stalemate. While in chess that means a tie, here it is a loss of the stalemated player. The next diagram shows a position where Black plays next. There are only two possible moves: either he captures a4 with b5 or advances b4. The first option leads to defeat since White replies b3:a4 and wins by stalemate. The second option leads to victory because White gets stalemated.

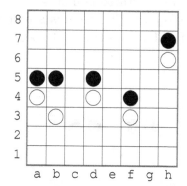

SanQi

This game was invented by L. Lynn Smith in 2003.

Materials

A hexagonal board with five hexagons per side, 25 white, 25 black and 25 grey stones.

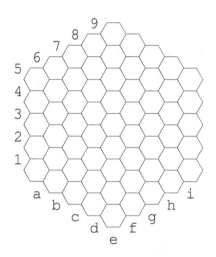

Rules

On each turn, each player may do one of the following actions:

a) Drop a stone of either color in an empty hexagon.

b) Replace a stone already on board with a stone of another color, but only if there are at least two more adjacent stones of the new color than the number of adjacent stones of the original color (e.g., a white stone adjacent to two white stones and four black stones can be replaced by a black stone).

Goal

The first player wins if he makes a ring of six stones of the same color. The second player wins if he makes a line with six or more stones of the same color. Also, any player may win by making a triangle of six stones of the same color.

The next diagram shows a ring, a line and a triangle pattern:

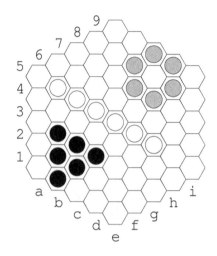

Notes

SanQi can be played on larger boards. (Players will need more stones.) The larger the board, the more strategic the game becomes.

The first player should focus on making rings while the second should focus on making lines. Triangles are usually used as threats since both players can win with a triangle. The hexagon enclosed by a ring may be occupied or not.

Players should avoid reducing their play to stones of just one color. If the adversary does that, then the other player should use the other two colors.

A stone can always change color unless it is protected by other adjacent stones. Of course, protection takes time. Players should try to take advantage of both stone protection and stone position. In the next diagram, the second player drops a black stone at i7. His goal is to create a majority around grey stone h7, in order to make a line of six black stones. The first player cannot resist this attack. Even if he drops a grey stone at g7 (around h7 are now two grey and three black stones), the second player is able to replace white h6 with a black stone. (The replacement is valid since there are three black stones and no white stone adjacent to h6.) After this reply, the first player cannot prevent the replacement of h6 at the next turn.

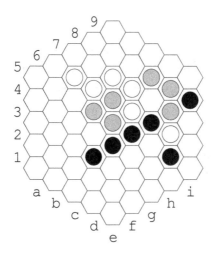

Players should not become too attached to a single structure. If a structure is threatened, there are always other places to play. A player can even use the adversary attack to start making a new pattern.

When making a ring or a line, players should try to prevent quasi-triangle patterns. The opponent may use them to destroy the other player's structure. Check the next position:

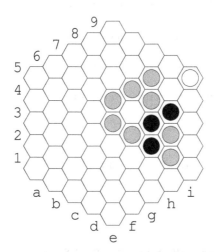

It is first player's turn. If he drops a grey stone at f6 (trying to win with a ring by threatening a replacement of the black stone at g5), he will lose the game since the second player can drop a grey stone at d5 and win with a grey triangle.

Semaphore

This game was invented by Alan Parr in 1998. Semaphore, despite the very small board, possesses an unexpected complexity, demanding attention from both players until almost the last moves.

Materials

A rectangular board with three rows and four columns, 12 green, 12 yellow, and 12 red stones.

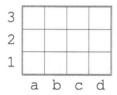

Rules

On each turn, each player must execute one of the following options:

1. Drop a green stone on an empty square.

2. Replace a green stone with a yellow stone.

3. Replace a yellow stone with a red stone.

Goal

The player that makes a three-in-a-row with stones of the same color wins the game.

Notes

Here, since this book does not use colors, green stones are represented as white, yellow as grey, and red as black.

The next diagram shows a position with three immediate wins: (i) replace the green stone at a3 (making a vertical line of yellow stones), or (ii) replace a yellow stone at d1 (making a diagonal line of red stones), or even (iii) replace the yellow stone at c3 (making a horizontal line of red stones).

112

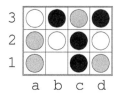

The game's outcome depends on the number of available moves until a winning pattern emerges. If that number is even, the next player wins, if odd, the next player loses. The problem (and interest) is that is not easy to evaluate that number of moves in an initial phase. The first player that deduces that number has a definitive advantage.

It is important to keep open as many options as possible. Players share all pieces, and so making a trap is always dangerous if the adversary understands it. The game is not a strategic one, but the short duration of each game and the simplicity of the rules make it a funny and fast game, easy to explain to children.

The next example shows an endgame. The board has only two moves until a winning pattern is made: (i) drop a green stone at b1, or (ii) replace the green stone at d2. That means the next player loses the game (if the other player is aware of these two moves, of course).

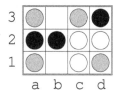

On the reduced 3×3 board (the tic-tac-toe board), there is a winning strategy for the first player. Drop a green stone at the center. The next player must replace it with a yellow stone. (Any other move means immediate defeat.) Then the first player replaces the yellow with a red stone. All other moves from the second player should be played symmetrically (around the center) until a winning move is available.

Slimetrail

This game was invented by Bill Taylor in 1992.

Materials

A hexagonal diamond board with ten hexagons per side, one black and 75 white stones.

Initial setup:

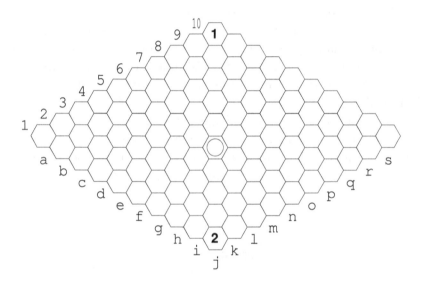

Rules

The white stone starts at j5.

On each turn, each player moves the white stone to an adjacent empty hexagon. After the move, the player drops a black stone at the white stone's initial hexagon. It is illegal to play in a way that prevents the white stone to reach at least one goal cell (j1 or j10).

Goal

The first player wins if the white stone moves to j10. The second player wins if the white stone moves to j1.

Notes

This is a very tactical game. It is essential for a player not to trap himself in a position where the adversary can cut all possible ways to the player's goal cell. If that happens, the game is already lost.

It is unimportant who moves the white stone to a goal cell. Even if the adversary moves the white stone to the player's goal cell, he loses the game. The next diagram shows a position where if the first player moves to k10, the second player must move to j10, thus losing the game.

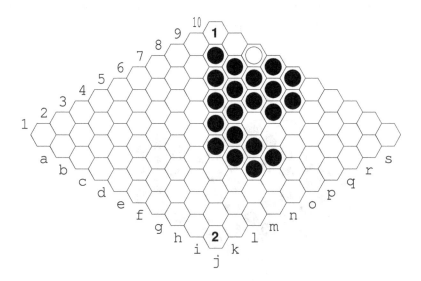

Handling how the white stone moves near the edges is very important. It is common to reach a position where one player can close all possible paths to the adversary goal cell, and that usually happens near the edges and corners, where the number of empty hexagons and adjacencies are smaller. In the next position, the second player moved to c2 leaving no good answers for the first player.

115

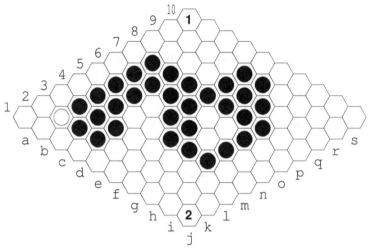

Notice that, once in the first line (a1 to j1), the second player can keep the white stone near the edge and move it to his goal cell. Let's observe the continuation of the previous position (the odd moves are from the first player):

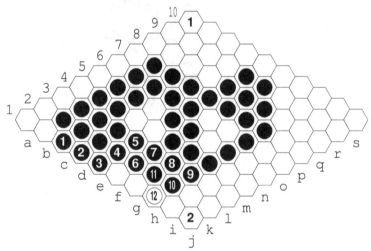

Now, the first player must move to i1, which provides an immediate win for the second player that moves to j1.

Variants

We can change the rules to allow players to move the white stone to a position where it is impossible to reach any goal cell. In this case, a player may also win by stalemating the adversary.

116

Another variant is to play Slimetrail in a square board, where the white stone can move to any orthogonally or diagonally adjacent empty square. There is more space to maneuver (eight adjacencies against six from the original game) which increases the tactical possibilities. Here is a complete game record:

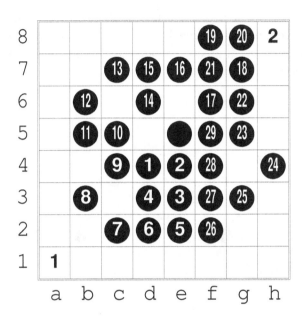

Some notes about this match. In move 9, the first player did not move to b2 (that would be easily handled by the adversary with c3) and kept an important square empty (which may be useful in the endgame). Between moves 10 and 12, we watch the second player try not to move away too soon from the adversary's goal cell. (He is trying to cut as many paths as possible.) At the same time, between moves 11 and 15, the first player tried to keep those paths open. If the first player can force the adversary to move to the empty left squares (columns a and b) he will win the game. With some difficulty, in move 18, the second player was able to move to g7 which was probably a bad move since the first player easily defended himself moving to f8. The insistence of move 20 was also quickly replied. Move 23 was decisive. The second player was unable to get the white piece to his goal cell; the remaining path was already too thin. With move 29, the second player resigned. Any further move will close the path to h8 (if g4 then h3).

Stooges

This game was invented in 2002 by Jorge Nuno Silva. It is an alignment game in which players try to get three-in-a-row in a small board, where it is legal to change the colors of the cells.

Materials

A diamond-shaped board, such as the one illustrated below. The squares of the boards have two sides, one black and one white. (Pieces of Othello/Reversi can be used.) Three white and three black pieces are also used. The initial position:

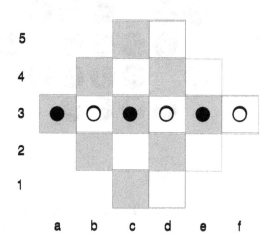

Rules

On his move, a player can do one of the following:

1. Move one of his pieces to an (orthogonally or diagonally) adjacent empty square of his color. White pieces only move on white squares and black pieces move on black squares.

2. Change the color of a square (from white to black or from black to white) provided the chosen square is empty and was not switched in the adversary's last move.

118

Restriction: If two consecutive moves consist of changing colors of squares, the next move must consist of moving a piece.

Goal

The winner is the player who creates a line (orthogonal or diagonal) with his three pieces.

Notes

This is a very tactical game. It seems easy to make a three-in-a-row, but, as the pieces must move only on same color squares, and these can be switched, it is wise to keep more than one possibility of movement for each piece.

This is a fast game where threats are used abundantly to keep the adversary from moving his pieces at his will. For example, in the situation illustrated, White threatens to win by playing c4 (this means to change the color of the square with coordinates c4), followed by d3-c4 (move the piece from d3 to c4). It is Black's turn.

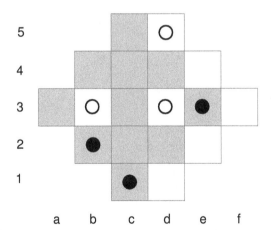

Black has a way out. The game unfolds: c1-c2; (threatening to win immediately with e3-d2) d2, c2-c3 (threatening e3-d4), d4, c3-c4 and the white threat is dismissed.

It is important to keep in mind the Ko-like rule: the last switch of color of a square cannot be undone. This rule is necessary to avoid an infinite cycle or color switches. Thus, a switch is a very strong winning threat.

That is the case of the following position: if White changes the color of d4, Black can do nothing since he cannot change it back or occupy it with a friendly piece. In his next move, White will make three-in-a-row and win.

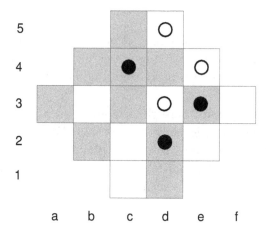

UN

This game was invented by João Neto in 2003. This game is similar to Slimetrail (See p. 114), since crossed hexagons cannot be used a second time.

Materials

A hexagonal board with five hexagons per side, one black and 50 white stones.

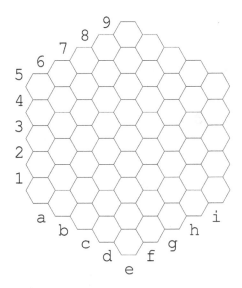

Rules

One player drops the black stone on an empty hexagon, and the other player decides who starts.

On each turn, each player slides the black stone over a line of one or more empty hexagons. In each crossed hexagon, the player drops white stones.

Goal

The player unable to move loses the game.

Notes

The next example shows the first ten moves of a UN game. One player dropped the black stone at h8. The other player decided to start and slide the stone to e8, placing three white stones at h8, g8, and f8. After that, the second player moved the stone to b5, placing another three white stones, and so on.

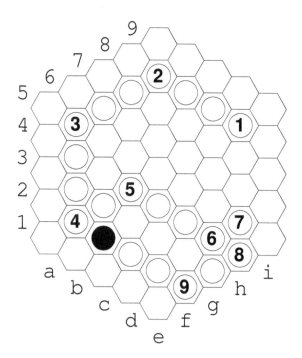

Considering the current position, the next player has two main options, (move to b1 or d3) both leading to defeat (if b1 then e1; if d3 then f3).

Usually, not all the hexagons are used in a game. The number of available hexagons always decreases, and it becomes easier and easier to move into a winning position. Let's observe the next play:

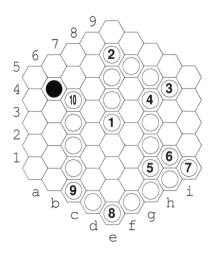

In this position, the current player moves the black stone to b1, which forces the next set of moves shown:

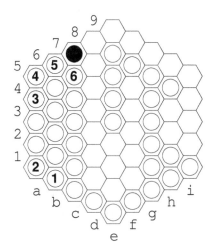

Now, the next player loses the game. If he moves to column d, the adversary replies to d2 and the game is lost. If, instead, he moves to e9, the next player has a winning sequence (h9, i9, i6, h6, h7).

This game can also be played on a square board (with orthogonal and diagonal slides). One possible extension is to use more than one black stone. Each player decides which black stone he wants to move on each turn. In this variant, the individual position of a single black stone is not so crucial for the final outcome of the game.

Y

This game was invented by Charles Titus and Craige Schensted. It is a connection game inspired by Hex (see p. 77), but where the goal is to connect all three edges.

Materials

A triangular board with eleven hexagons per side, 30 white and 30 black stones.

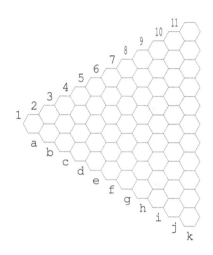

Definitions

Group — a connected group of pieces of the same color.
Y — a group that connects all three edges.

Rules

On each turn, each player drops a friendly stone on an empty hexagon.

After the first drop, the second player has the option to swap colors (i.e., he becomes the first player, and the adversary continues the game as the second player).

Goal

The player that achieves a Y wins the game.

124

Notes

The next diagram shows a Y:

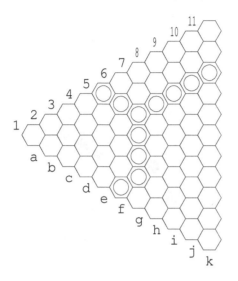

A corner belongs to two edges, and they can be used to construct Y structures. The next group is also a Y:

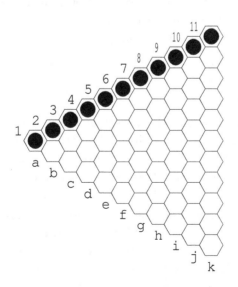

The next example shows an endgame, where Black has a definitive advantage, regardless who is next. Any White attack can be neutralized. For example, if White drops a stone at d2, Black replies e2, making his virtual connection into an effective connection.

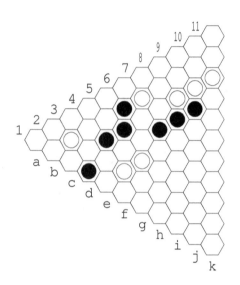

The main strategy, like in most connection games, consists in keeping as many connective options as possible and preventing the adversary from doing the same. Most of what was said about Hex can be said about Y.

Variants

A nice variant is to apply the progressive mutator to Y. So, the first player in the first turn makes one drop, then the second player makes two drops, then three, four, ... There is, however, a restriction to prevent the game from becoming trivial: no two stones dropped in the same turn can belong to the same group!

The next diagram shows the first four moves on a progressive Y game. Black started. In the second turn, White plays already four stones. Stones h2 and j4 create a double connecting threat with h4. Black cannot stop both in the same turn (that is, playing in h3 and i4) because these two stones would belong to the same group.

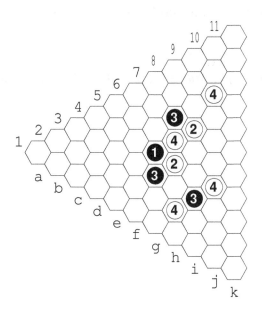

We get another (slower) variant allowing each player to drop just two stones (except in the first move, where the first player just drops one stone). There is the same restriction (no stones in the same group at the same turn). The next diagram shows the initial moves of a game with these rules:

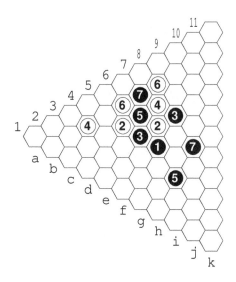

After the fourth black move, his positional advantage is already noticeable .

Reference

Schensted, C. & Titus, C., *Mudcrack Y & Poly-Y*, Neo Press, 1975.

Chapter 3

Nim Games

Combinatorial Impartial Games

This chapter focuses on a type of game that has a very rich mathematical history. We present the main results, which are useful to play the games. We provide the interested reader with the bibliographical references ([CON, BCG]).

These games, also known as Nim games, are characterized by the following conditions:

1. There are two players.

2. The set of possible positions of the game is well understood.

3. In each position, the moves available are the same for both players (there are no White and Black, or any other way of telling the players apart).

4. Players alternate.

5. The game ends when one of the players is faced with a position where no legal move is possible.

6. Whoever plays last wins. This means that the loser is the first person unable to play.

7. The game takes a finite number of turns, independent of the way the play goes.

Let us start with an example. Consider a pile of 17 beans. Two players alternate taking beans from the pile. The rules say that, in each move, a player can choose to take one, two, or three beans from the pile. Whoever takes the last bean wins.

To understand completely this game, we must do some retrograde analysis, starting from the final (empty) position. If there are one, two, or three beans, the player to move wins taking them all. If there are four, the situation changes with the player to move able to take one (leaving three), two (leaving two), or three beans (leaving one). In any case, he will lose because, as we just saw, the player to move wins when facing a pile with one, two, or three beans. The player that leaves four beans will win the game. In a similar way, we can see that it is a good idea to leave eight, twelve, or sixteen beans for your opponent.

These positions that give, under perfect play, the victory for the player who leaves them, are called *P-positions*. The remaining positions, which are winning positions for the player to move, are called *N-positions*.[1] The set of all P-positions is represented by $\mathcal{P}$ and the set of all the N-positions by $\mathcal{N}$.

---

[1] The letters P and N are the initials of *Previous* and *Next*.

In our example, the P-positions are $0, 4, 8, 12, 16$. We then have

$$\mathcal{P} = \{0, 4, 8, 12, 16\}, \quad \mathcal{N} = \{1, 2, 3, 5, 6, 7, 9, 10, 11, 13, 14, 15, 17\}.$$

If, instead of 17, we had a larger number of beans, it was easy to see that the P-positions would correspond to multiples of 4.

Given a combinatorial impartial game, we can always start working from the final positions and figure out which positions are in $\mathcal{P}$ and which are in $\mathcal{N}$. However, sometimes this work gets very complex. The basic algorithm as the follows:

1. All terminal positions are P-positions.

2. All positions from which a player can move to a P-position are N-positions.

3. All positions from which one can only get to N-positions are P-positions.

4. If step 3 introduces no new positions, the algorithm halts; otherwise, go to step 2.

Another way of referring to this partition of the positions of a game is the following:

Characterizations of $\mathcal{P}$ and $\mathcal{N}$. The P- and N-positions are defined recursively by the conditions:

i) All the terminal positions are P-positions.

ii) From any N-position, there is (at least) one move taking to a P-position.

iii) From any P-position, all moves lead to N-positions.

Subtraction Games

From any set of integers $\{1, 2, 3, 4, 5, \ldots, n\}$ (beans, for instance) we can define a game similar to the one we analyzed previously, changing the number of beans each player can take in a move. The game we treated corresponds to a set of legal subtractions given by $\{1, 2, 3\}$ and can be represented by $S_{\{1,2,3\}}$. (As it began with 17 beans, we could be more precise and write $S_{\{1,2,3\}}(17)$).

We can analyze more general subtraction games. Consider, for instance, the game $S_{\{2,3\}}$.

The terminal positions are $0, 1$, which are P-positions. The positions from which we get to a terminal position are $2, 3, 4$; they must be labelled N. From 5 or 6, a player can only get to $2, 3, 4$. Therefore, 5 and 6 are P-positions. The pattern is clear.

We have

$$\begin{array}{cccccccccc}
0 & 1 & 2 & 3 & 4 & 5 & 6 & 7 & 8 & 9 & \ldots \\
P & P & N & N & N & P & P & N & N & N & \ldots
\end{array}$$

Other games can be analyzed this way, even board games. The following pages contain some games of this kind. We invite the reader to identify the P- and N-positions and, consequently, master the games.

White Knight

In this game, a checkered board is used, such as a chessboard. At the beginning, there is a single chess piece on the board, a white knight. Each move consists of moving the knight according to the rules of chess, but restricted to the directions shown below. When it gets to one of the shaded squares, a terminal position is reached; no more moves are possible. The player that moved the knight to one of those squares is the winner.

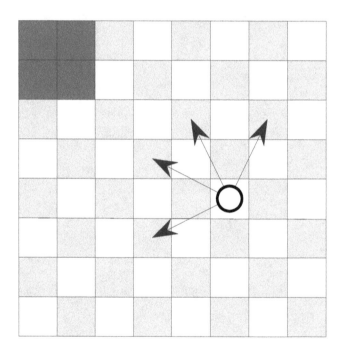

Reference

E. Berlekamp, J. Conway, R. Guy, *Winning Ways*, A. K. Peters.

Wyt Queen

At the beginning there is a white queen in a rectangular board, such as a chessboard. Each move consists in moving this piece according to the usual chess rules with the restriction that it must move north, west, or northwest. The figure below illustrates the movements. The players play alternately until the queen reaches the top left corner. The player that puts the queen there is the winner.

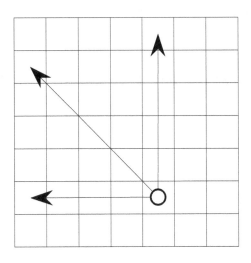

Reference

E. Berlekamp, J. Conway, R. Guy, *Winning Ways*, A. K. Peters.

Whithoff

This game starts with two piles of beans. Each move consists in one of the following possibilities:

1. Choose one of the piles and decrease it (take at least one bean; can take the whole pile).

2. Take the same number of beans from both piles.

Whoever takes the last bean wins.

Reference

E. Berlekamp, J. Conway, R. Guy, *Winning Ways*, A. K. Peters.

JIL

This game, invented by Jorge Nuno Silva, happens in a n-tuple of nonnegative integers. The players alternate according to the following rule: a move consists of choosing a component of the n-tuple, subtracting one unity from it, and adding nonnegative quantities to the higher-order components. (In an n-tuple the order of the components increases from left to right.)

The first player that finds all the components of the n-tuple zeroed, on his turn loses.

Let us look at an example. For $n = 5$ and initial position $(10, 7, 10^{10}, 0, 2^{2^2})$ we have, for example, this sequence of legal moves:

$$(10, 7, 10^{10}, 0, 2^{2^2}) \mapsto (10, 6, 10^{10^3}, 99^{99}, 2^{2^2}) \mapsto (10, 6, 10^{10^3}, 99^{99}, 2^{2^2} - 1).$$

Is this game finite, independent of the original setup? The question is, given any initial n-tuple, can we be sure that the game will end after finitely many turns?

Assuming it does end in finite time, what is the winning strategy (to identify $\mathcal{P}$ and $\mathcal{N}$)?

For instance, from the position

$$(22, 222, 2, 3, 2^3, 10),$$

who wins: the first or the second player?

References

Silva, J. N., "Notas sobre o Problema anterior e JIL-Jogo Indefinidamente Longo," *Boletim da Sociedade Portuguesa de Matemática*, 43, October 2000: 143–147.

Silva, J. N., "Notas sobre o Problema anterior e Exponenciação Comutativa," *Boletim da Sociedade Portuguesa de Matemática*, 44, April 2001: 119–121.

LIM

In this game, invented by Jorge Nuno Silva, we use three piles of beans.

Each move consists of choosing two piles, taking the same number of beans from each of them, and adding the same number to the third one. The player who, on his turn, finds two empty piles, loses.

Here are some examples of legal moves:

$$(5, 3, 7) \mapsto (2, 6, 4) \mapsto (4, 4, 2).$$

Or, in a diagram:

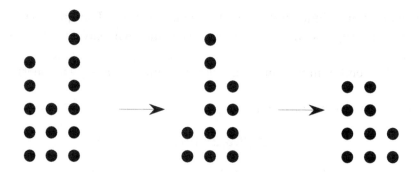

References

Silva, J. N., "Notas sobre o Problema anterior e LIM," *Boletim da Sociedade Portuguesa de Matemática*, 46, April 2002: 119–124.

Silva, J. N., "Notas sobre o Problema anterior e Corpo Estranho," *Boletim da Sociedade Portuguesa de Matemática*, 47, October 2002: 97–100.

Nim

This classic game was the first combinatorial game to deserve the attention of a professional mathematician, Charles Bouton, who solved it and presented its theory in a 1902 paper ([BOU]).

At the beginning there are some piles of beans. Each player, on his turn, chooses a pile and takes some beans from it (at least one, he can take the whole pile).

Whoever takes the last bean is the winner.

If there is only one pile of beans, the analysis is very easy. For any nonempty pile, the player to move should take all the beans and win. If the pile is empty, the previous player has won already.

One-pile Nim is too easy:

Number of beans	0	1	2	3	4	...
Type	P	N	N	N	N	...

Two-pile Nim isn't too difficult either. If the piles have different numbers of beans, the player to move should make them equal, and, from then on, imitate his adversary's move. For example, if the two piles have 4 and 6 beans (we can represent them by the ordered pair $(4, 6)$) the good move is for $(4, 4)$. Now, what the adversary does to one pile is also done by the player to the other pile, leaving equally sized piles again . This kind of strategy, replicating the adversary's moves in another component of the game, is known as *Tweedledum and Tweedledee*.

The characterization of the positions in two-pile Nim:

$$(n, m) \quad \text{is} \quad \begin{cases} \text{P} & \text{if} \quad n = m \\ \text{N} & \text{if} \quad n \neq m \end{cases}$$

Three-pile Nim is not as easy as the previous cases. We need some new concepts to fully analyze it.

We obtain the *Nim-sum* of two nonnegative integers x, y, representing them in base 2 and adding their coefficients modulo 2 (that is, $0 + 0 = 0$, $0 + 1 = 1$, $1 + 0 = 1$, $1 + 1 = 0$). We use the notation $x \oplus y$.

Consider, for example, $5 \oplus 7$. We have $5 = 2^2 + 1$, $7 = 2^2 + 2 + 1$. Therefore, in base 2,

$$5 = (101)_2 \qquad 7 = (111)_2$$

Adding the coefficients modulo 2, we obtain

$$\begin{array}{r} 1 \ 0 \ 1 \\ + \ 1 \ 1 \ 1 \\ \hline 0 \ 1 \ 0 \end{array}$$

Therefore, as $(010)_2 = 2$, we got $5 \oplus 7 = 2$.

Nim-sum has nice properties:

1. Associativity: $x \oplus (y \oplus z) = (x \oplus y) \oplus z$;
2. Commutativity: $x \oplus y = y \oplus x$;
3. 0 is neutral: $0 \oplus x = x$;
4. Each number is its own additive inverse: $x \oplus x = 0$;
5. The cancellation law holds: If $x \oplus y = z \oplus y$, then $x = z$.

This operation is important because Bouton characterized the positions in the general Nim game in terms of the Nim-sum of the sizes of the piles of beans. If we represent a position with n piles of beans with $x_1, \ldots, x_n$ beans by $(x_1, \ldots, x_n)$, we can state the main result as follows.

Bouton's Theorem: The position $(x_1, \ldots, x_n)$ is a P-position if and only if $x_1 \oplus \cdots \oplus x_n = 0$.

Note that this theorem holds also for the one- and two-pile Nim, which were analyzed before.

Let us consider an example.

Consider the Nim game with four piles $(3, 5, 7, 9)$. We must calculate $3 \oplus 5 \oplus 7 \oplus 9$. As $3 = (11)_2$, $5 = (101)_2$, $7 = (111)_2$, $9 = (1001)_2$, we have

$$\begin{array}{r} 0 \ 0 \ 1 \ 1 \\ 0 \ 1 \ 0 \ 1 \\ 0 \ 1 \ 1 \ 1 \\ + \ 1 \ 0 \ 0 \ 1 \\ \hline 1 \ 0 \ 0 \ 0 \end{array}$$

thus, $3 \oplus 5 \oplus 7 \oplus 9 = 8$. The given position is an N-position. The (unique) winning move consists of taking 8 beans from the 9 pile.

Check $3 \oplus 5 \oplus 7 \oplus 1 = 0$:

$$\begin{array}{r} 0 \ 0 \ 1 \ 1 \\ 0 \ 1 \ 0 \ 1 \\ 0 \ 1 \ 1 \ 1 \\ + \ 0 \ 0 \ 0 \ 1 \\ \hline 0 \ 0 \ 0 \ 0 \end{array}$$

We just presented the theory of Nim with some detail because it is important for lots of other games. If you can play Nim well, then you can play hundreds of other games well. Some of the following are just versions of Nim. So this theory applies to them. However, sometimes the identification of a given game with a Nim variant is not clear.

Plainim

We use a chessboard. Some pieces, all alike, are randomly distributed by the squares. Each square can hold at most one piece. A move consists of choosing one occupied square taking its piece out of the board. In the same turn, the player can change the status of each of the squares in the same row, to the left of the previously chosen (put pieces, take pieces or neither).

Whoever takes the last piece wins.

We show an example of a legal move (in the last row, taking a piece and putting two).

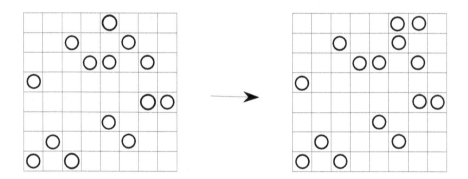

Notes

Identifying each empty square with 0 and each occupied one with 1, we obtain the binary representation of a nonnegative integer in each row, .

The analogy with Nim is now clear.

Reference

E. Berlekamp, J. Conway, R. Guy, *Winning Ways*, A. K. Peters.

Nimble

This game occurs in a strip $1 \times n$. At the beginning, there are some coins randomly distributed by the cells. Each cell can contain any number of coins. A move consists of choosing a coin and moving it to any cell to the left of the original position. Jumping over coins is permitted. The game ends when all coins are in the leftmost cell. The last player to move is the winner.

Notes

For each coin consider its distance to the leftmost cell of the strip. We obtain a set of numbers this way . Each move consists in decreasing one of them.

If we take a Nim game where the sizes of the piles are defined by these numbers, we get a similar game. How is it similar to Nimble?

Reference

E. Berlekamp, J. Conway, R. Guy, *Winning Ways*, A. K. Peters.

Turning Turtles

The game was created by the Berkeley mathematician H. Lenstra. A row of n coins is placed on a table. Some show their **F**aces and the others show the **V**ersos. For instance, with $n = 13$, we could have

$$V \quad F \quad V \quad V \quad F \quad V \quad V \quad V \quad F \quad F \quad V \quad F \quad V$$

Each move consists in choosing a coin that shows F, turning it, and, optionally, changing the state of some coins to the left of the one chosen before. Whoever turns the last F to V wins.

An equivalent implementation uses checker pieces (white = F, black = V). Here a move consists of changing a white for a black and switching the color of some piece to the left of the one previously chosen. The game ends when there are only black pieces and the last player to move wins.

The position corresponding to the situation above:

Notes

It is natural to identify, for each F, a Nim pile with the number of beans given by the distance of F to the leftmost coin. However, this does not work, since a move in Nim consists only in decreasing one of the present numbers.

Maybe we should recall that in that strange "sum" we introduced before (see p. 139), addition sometimes looks like subtraction ($x \oplus x = 0$).

Reference

E. Berlekamp, J. Conway, R. Guy, *Winning Ways*, A. K. Peters.

Silver Dollar

This game occurs in a strip $1 \times n$. At the beginning, there are some coins randomly distributed in the cells, at most one in each cell. Each move consists of choosing a coin and moving it any number of cells to the left, but it cannot jump over occupied cells. The game ends when the coins are all jammed on the left part of the strip. The last player to move wins.

Notes

Identifying a coin with the distance to the leftmost cell does not work because jumping over other coins is not allowed.

Hint: Identify, alternately, from right to left, each space between consecutive coins with a Nim pile of a matching number of beans. We obtain a Nim game (in our example with piles of 1, 0, 0, 1 beans). Each move decreases one of these numbers. It can increase others, but that is not strategically important.

Reference

E. Berlekamp, J. Conway, R. Guy, *Winning Ways*, A. K. Peters.

Stairs

There are n steps in these stairs. Each step contains a nonnegative number of pieces. Each move consists in passing some of the pieces from one step to the one immediately below. The game ends when all the pieces are in the bottom step, the last player being the winner.

Another implementation consists of using a row of piles of pieces, each move consisting of taking some pieces from one pile to its left neighbor.

An n-tuple of nonnegative integers can represent the situation. Accordingly, the move $(7, 2, 1, 4, 5, 1) \mapsto (7, 2, 3, 2, 5, 1)$ corresponds to the original move below.

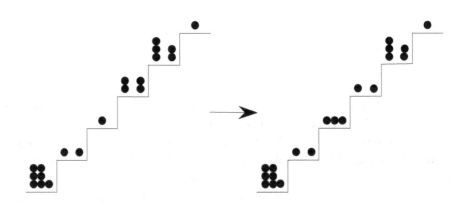

Notes

Label the steps, from left to right, with the positive integers. Consider the Nim game associated with the even numbered steps. Each move on those steps corresponds to take some beans from one of the piles. Some moves can increase these numbers, but they are not strategically important. (Why?)

Northcott

This game uses a rectangular board $n \times m$. There are two players: White and Black. In each column, there is one white piece and one black piece. Each move consists in moving a friend piece forwards or backwards any number of squares as long as it does not leave its column or jump over any other piece.

A player that cannot move, because all his pieces are trapped against the border, loses.

A possible initial position:

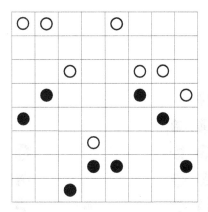

Notes

Consider the eight-pile Nim game with one pile of beans for each board column, the size of each pile being given by the number of squares between the two stones in the corresponding column.

Northcott looks like this game of Nim. However, as there are some moves that increase the number of squares between two stones of the same column, Northcott has some extra possibilities. It turns out that the extra moves are finite and not relevant for the winning strategy. (Why?)

The existence of two colors does not violate the principle of p. 131, because the only relevant parameter is the number of free squares between pieces.

Reference

E. Berlekamp, J. Conway, R. Guy, *Winning Ways*, A. K. Peters.

Nim$_k$

This game, created by H. Moore, uses n piles of beans. Each move consists of taking beans from at most k piles with $k < n$. At least one bean from one pile must be taken, and the quantities taken from the piles can be different. Whoever takes the last bean wins.

For $k = 1$, we have the usual Nim game.

One example of a legal move, where $n = 5, k = 3$, is:

$$(5, 4, 4, 7, 6) \mapsto (5, 0, 3, 7, 2).$$

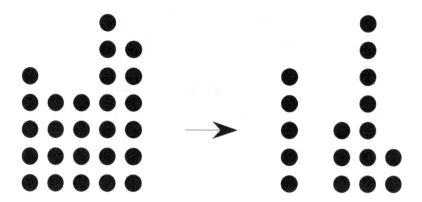

Notes

Binary representation and modulo 2 addition played a role in the understanding of Nim. Maybe an adaptation of that approach works here.

Reference

Gardner, M., *Knotted Doughnuts and Other Mathematical Entertainments*, Freeman and Company, 1986.

Blocking Nim

This game is played with piles of beans, just like Nim. The only difference is that before each move, the adversary states, for each column, a quantity of beans that cannot be subtracted from that column.

For instance, the position corresponding to piles of $1, 3, 5, 6, 2$ beans can be represented by

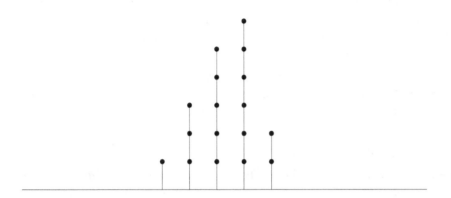

Notes

The Nim sum of the sizes of the piles is the main parameter to decide on the optimal restrictions to impose on the adversary.

Reference

George Washington University Problems Group, "Blocking Nim," *College Mathematics Journal*, vol. 35, no. 5, November 2002: 414–415.

Games on Graphs

In this section, we will learn how to represent some games using graphs. That will make possible to get a classification of the positions which is better than the P-N characterization.

We begin by stating some definitions.

A digraph, or directed graph, is a system containing one set of *vertices* (the positions of the game) X and one map F that assigns a subset of X to each element of X, x, $(x \mapsto F(x) \subset X)$. We represent it by (X, F). $F(x)$ is the subset of positions for which a player can move from x. If $F(x)$ is empty, we say that x is *terminal*.

A game in a graph is defined by picking a vertex to be the initial position, x_0, and stating the rule that the players alternate choosing, for each vertex x, another in $F(x)$. When a player, in his turn, has to move from y with $F(y)$ empty, he loses.

We consider only finite games.

The games on graphs can be analyzed with a technique similar to the one used to get the characterization in P- and N-positions in the impartial games, starting at the terminal vertices.

However, we introduce a function that gives us some extra information, and will be important in what follows.

Recall that, given a set of nonnegative integers A, we represent by mex A the smallest integer not in A. For example, mex $\{0, 1, 3, 44\} = 2$.

Grundy function of a graph (X, F) is the function $g : X \to \mathbb{N}_0$ defined by[2]

$$g(x) = \text{mex} \left\{ g(y) \, : \, y \in F(x) \right\}.$$

Note that we find the values of the Grundy function recursively. For each x, the value of $g(x)$ depends on the values that $g(y)$ assumes when y runs through $F(x)$. We should then start with the terminal vertices, x, for which $F(x)$ is empty, giving, in this case, $g(x) = 0$.

Let us work through an easy example.

---

[2]We represent the set of nonnegative integers by $\mathbb{N}_0$.

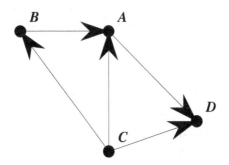

D is a terminal vertex, therefore $g(D) = 0$. No other vertex is terminal. As $F(A) = \{D\}$, and we know $g(D)$, we can calculate $g(A)$. We have

$$g(A) = \operatorname{mex}\{g(D)\} = \operatorname{mex}\{0\} = 1.$$

We have now $g(B) = \operatorname{mex}\{g(A)\} = \operatorname{mex}\{1\} = 0$. Finally,

$$g(C) = \operatorname{mex}\{g(A), g(B), g(D)\} = \operatorname{mex}\{0, 1\} = 2.$$

Labelling each vertex with its Grundy value:

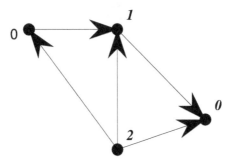

Note that, in a Nim game, the Grundy function of a pile coincides with its size.

Consider again the subtraction game $S_{\{2,3\}}$. We do not need to draw the graph to figure out the Grundy values in the positions of the game. The terminal vertices correspond to $0, 1$. Therefore, $g(0) = g(1) = 0$.

We have $F(2) = \{0\}$, and thus, $g(2) = \operatorname{mex}\{0\} = 1$. In a similar way, we conclude that $g(3) = 1$.

As $F(4) = \{1,2\}$, we get $g(4) = \text{mex}\{g(1), g(2)\} = \text{mex}\{0,1\} = 2$. As $F(5) = \{2,3\}$, we get $g(5) = \text{mex}\{g(2), g(3)\} = \text{mex}\{1\} = 0$. The pattern becomes clear:

$$
\begin{array}{ccccccccccccc}
x & 0 & 1 & 2 & 3 & 4 & 5 & 6 & 7 & 8 & 9 & \ldots \\
g(x) & 0 & 0 & 1 & 1 & 2 & 0 & 0 & 1 & 1 & 2 & \ldots
\end{array}
$$

Comparing with the previous anlysis of the game, we see that the P-positions are exactly those for which the Grundy function vanishes. This is general, as we can easily deduce from the very definition of Grundy function and the characterization of $\mathcal{P}$ and $\mathcal{N}$:

1. $g(x) = 0$ if x is terminal;

2. If $g(x) = 0$ and $y \in F(x)$, then $g(y) \neq 0$;

3. If $g(x) \neq 0$, then, for some $z \in F(x)$, we have $g(z) = 0$.

Sums of Impartial Games

One of the possible ways of playing several games at the same time is the following. If we have n games, a player, in his turn, chooses one of them and plays a legal move in it, leaving the others untouched. The players alternate until terminal positions are attained in all the games. Whoever makes the last move wins. This new game is called the *sum* of the original games.

This is one way of building new games from old ones. Often, even if the summands are easy games, their sum gets very complex, and difficult to analyse. One example: regular Nim is just the sum of several one pile Nim games.

However, there is a result that makes it easy to find the Grundy function of the sum, if we know the Grundy function of the summands. We state it here in the language of graph theory. Consider games $(X_1, F_1), \ldots, (X_n, F_n)$, with corresponding Grundy functions $g_1, \ldots, g_n$, respectively. Each position in the sum of these games can be identified with an n-tuple of positions of the summand games, $(x_1, \ldots, x_n)$, where $x_1 \in X_1, \ldots, x_n \in X_n$. Then, the Grundy function of the sum is

$$g(x_1, \ldots, x_n) = g_1(x_1) \oplus \cdots \oplus g_n(x_n).$$

As we know already that the P-positions for the sum are those that have a Grundy value of zero, it is enough to know the Grundy functions of the components and their Nim-sum to establish an optimal strategy for the sum.

For example, we analyze the game we get when we sum a game, J_1, which is three-pile Nim where the sizes are 3, 5, and 7 with J_2, the subtraction game $S_{\{2,3\}}(4)$.

To calculate the value of the Grundy function for the Nim component: $3 \oplus 5 \oplus 7 = 1$. In the game $S_{\{2,3\}}(4)$, a value of 2 corresponds to one pile of 4 beans, as we saw on page 133. Therefore, the value $1 \oplus 2 = 3$ corresponds to the sum of these games. It is an N-position. A good move consists in taking 2 beans from the pile corresponding to the game $S_{\{2,3\}}(4)$ since, in this game, a pile with 2 beans has a Grundy value of 1 and $1 \oplus 1 = 0$. We would get to the position below:

Was this winning move unique?

Using other impartial games, the reader can build lots of new games, using the previous definitions. One such an example is on the next page.

Wyt Queens

This game uses a rectangular board, for instance a chessboard. The pieces are all alike, white queens, which move as in chess, but only in the directions north, west or northwest. Each move consists of moving one of the queens any number of squares. Each square can hold more than one queen. Whoever puts the last queen in the top left corner wins.

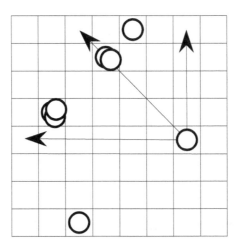

Notes

To each queen we can associate a Wyt Queen game (see p. 135). Our game is then the sum of these games.

Reference

E. Berlekamp, J. Conway, R. Guy, *Winning Ways*, A. K. Peters.

Green Hackenbush

To play this game we draw a bush, that is, a set of segments connected to the *ground*. Each move consists of erasing one segment, but all the other segments that lose connection to the ground must disappear also. Whoever cuts the last segment wins.

The easiest case corresponds to having several stalks:

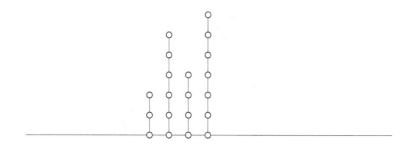

This is just a Nim game with piles of sizes 2, 5, 3, and 6.

In general, the situation is more complicated.

An example of a legal move:

If a player decides to cut the segment **a**, then three other segments go away.

156

This is, of course, an impartial game. According to our theory, each of its positions corresponds to a Nim pile with a size given by the Grundy function.

It is not true that each particular bush is the sum of smaller bushes, because these are not usually isolated. Therefore, the theory we presented on the sums of games does not apply here. However, we'll present two principles particulary useful for calculating the Grundy values of the bush's positions.

The simplest bushes are *trees*. These are bushes that have one root. They are easy to spot since in a tree there is only one way from each vertex to the ground:

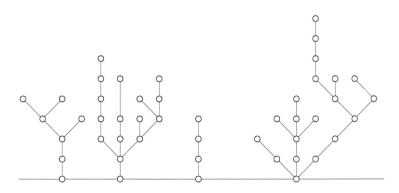

To analyze this kind of "plant" we have the *Colon principle*: if n stalks converge in a vertex, they can be replaced by one, the size of which is given by the Nim sum of the sizes of the original stalks.

This principle is applied from the most remote segments towards the root, simplifying in each step, ending with just one stalk. Recall that the size of a stalk gives its Grundy number.

An example:

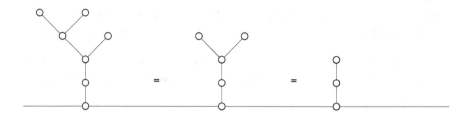

because $1 \oplus 1 = 0$.

Still another example:

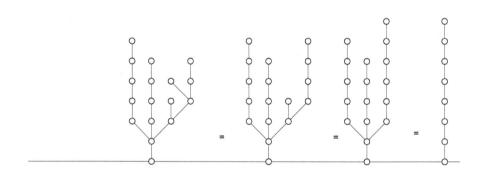

because $1 \oplus 2 = 3$, $1 \oplus 4 = 5$ and $4 \oplus 5 \oplus 6 = 7$.

For the general case, we'll allow more than one root and the existence of cycles (closed paths in the bush) and loops (curved segments, connecting a vertex to itself). As far as playing the game is concerned, loops and regular segments are indistinguishable. We can put a segment in the place of a loop anytime we want.

We'll be dealing with bushes like these:

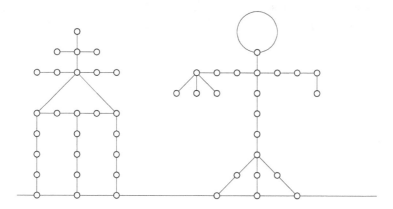

Note also that the ground behaves like a special vertex.

To *fuse* two adjacent vertices means to make them coincide. The segment connecting these vertices is transformed in a loop.

To analyze complex bushes, we use the *fusion principle*, which states that, in any bush, each cycle can be fused into one of its vertices without changing the global Grundy.

An example, where we also use that fact that the ground is just a special vertex:

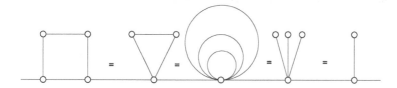

In the first step, we identified the ground with a vertex. In the second, we applied the fusion principle to some vertices. In the third, we put segments in the place of the loops. The last step is just an application of the Colon principle.

Successive uses of these two principles make possible the determination of the Grundy value of very complicated bushes. A last example is given below:

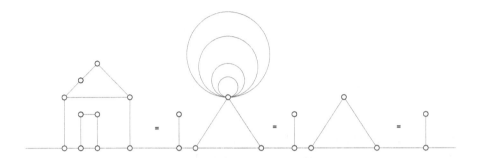

We knew already that the door is equivalent to a segment. The roof disappears by fusion of its vertices followed by the Nim addition of its segments $(1 \oplus 1 \oplus 1 \oplus 1 = 0)$. Two segments, which add up to zero, and an isolated one are all that is left. Therefore, the whole picture has a Grundy value 1. It is then a N-position.

This is a game fit for pencil-and-paper (or chalk and blackboard). The reader is invited to draw very messy forests and start playing them.

Reference

E. Berlekamp, J. Conway, R. Guy, *Winning Ways*, A. K. Peters.

Chapter 4

Games for Three

Games for Three Players[1]

Between family and friends, it is common to make a group around a table to play board games. In this age of e-mail and online servers of various sorts, it is becoming increasingly easier and more common for many players to join together to play games via electronic media. Most abstract games are for two players. Some, especially chess variants, are extended to four players. Very few have good three player versions. Why is that?

Moving from two-player games to multiplayer games creates a social environment, which allows alliances, threats, betrayal, and a raft of group behaviors that go well beyond the mere dictates of the rules. Some games, which would have no interest with just two players, may flourish with several players; however, the opposite is also true for some games.

We are referring to games where the players may interact with each other. (This excludes race games.) In these games the main problem is the management of alliance scenarios. Usually, the game designer creates a game with an even number of players (the most common number being four) and sometimes specifically defines the alliances in the rule-set. This is very common in card games or four-player chess games. When choosing an odd number of players, which is almost a guarantee of constant game tension, games tend to have five or seven players (like Diplomacy) to smooth one potential dilemma: the *petty diplomacy problem* (PDP).

PDP is a feature that can destroy many promising games. Although in the form commonly known as "the Tall Poppy effect," it is almost inevitable that as soon as one player gets ahead, the others cut him down until the victim is weak enough and the ganging up shifts to the next stronger player. Less pleasantly, PDP sometimes produces behavior called "the kingmaker effect": when a player is about to lose, he uses his remaining strength in a murder/suicide strategy to defeat another player. If the number of players is three, then defeating one player means granting victory to the other. So in a three-player game, king-making is a powerful and nasty bargaining tool. Also, with such a small set of players, the suicide strategy may not be entirely effective and the attacked player may become so weak as to attempt to inflict the suicide/murder pact on another player. The cycle continues and the game dynamic is compromised.

Worse still, PDP opens up the games to problems of spite and revenge. There's one article ([STR]) relating three-player Hex which states the Mc-

---

[1] This chapter is based on an article written by João Neto, Bill Taylor and Cameron Browne.

Carthy Revenge Rule (or guideline): "If I am about to lose, I will inflict as much damage as possible on the player who put me in this position." This balancing mechanism is widely used and has a rough justice about it; however, it is difficult to enforce in practice. In some cases, both opponents are equally to blame for a player's disastrous position. Such tendencies may play a vital part in simulation board games such as Diplomacy and Monopoly. However, we strongly feel that such off-the-board bargaining should have no place in abstract board games. Of course, friendly reminders from one player to another, that the third is about to become unassailable unless a certain action is taken, are probably unavoidable and should not be legislated against. Nonetheless, the open use of threats and revenge to induce such action is most unpalatable.

We have thought about this subject and tried several ways to inhibit PDP with augmented rule-sets, keeping in mind that it may be impossible to eliminate it completely. It's possible that complete elimination of PDP may not be achieved and not even desirable (e.g., pure race games do not have PDP but tend to be intellectually boring games). So PDP may well add new layers and worthwhile complexity to the play. A good rule of thumb seems to be "Tall Poppy good, threats and revenge bad."

In other words, the ability to form micro-alliances (a PDP effect) is not only a side effect of three-player games; it may well be crucial to their success. Some designers such as Wayne Schmittberger go to great pains to reduce micro-alliances in games. But perhaps this is not a good thing, and such micro-alliances should be encouraged! But that's a thin line. Too much PDP and the game might cycle unless it has a specific length. If not, as in many games, it may never end or just produce endless stalemates (as in Diplomacy).

The basic idea that we adopt, is that no player may allow the next player to make an *immediate* win on his next move. This does not, of course, prevent him from making a more subtle move that virtually guarantees one of the others will win later on but does at least remove the grosser forms of PDP and spitefulness. We call this rule, or strictly speaking meta-rule, the STOP-NEXT rule.

The idea of an immediate win (that is, the fact that the next turn may be the last one legally played) is more clearcut in some games than others. For some games, we have found it wise to extend this rule a little (e.g., if the following series of moves is virtually forced) or even reduce it a little (e.g., if it is hard to tell at a glance that a won position is indeed won). But usually there is no problem. We assume this rule is in place in all of our following games. Further clarification is provided if needed.

164

We now present the rules of the most promising games that we have recently studied. Board sizes are not fixed and can be adjusted to taste.

Throughout the text we use the following definitions:

MOVER: the player in the process of moving or who has just moved.

NEXT: the next player to play.

OTHER: the remaining player.

Shared pieces

One way to reduce PDP is to use shared pieces. Here, in a sense, everyone is helping everyone else, even in the two-vs-one scenarios. This next game (by Bill Taylor) uses just two types of stones for three players, in a hexagonal connection game.

Triskelion

This game is played on the following board with 30 white and 30 black stones:

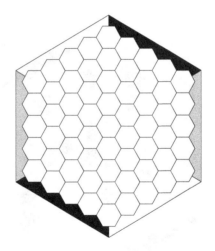

Each player owns an opposite pair of sides.

Rules

On each turn, each player drops one stone on an empty hexagon. The color of that piece must be different from the stone dropped by the previous player (i.e., stone colors alternate). The first stone to be dropped is black.

Goal

If two opposite sides are connected by a single color, the owner wins. (If this happens simultaneously, MOVER wins if he was one of the owners; otherwise, NEXT wins). If a Y pattern is made connecting 3 non-adjacent edges with one color, MOVER wins.

Notes

Notice that the number of piece types (2) and number of players (3) are out of phase every second turn. This means that if a player plays a black stone at turn N, he will only play another black stone at turn $N+2$. Thus, the players must constantly focus on chains of different colors.

This simple method of alternating both three players and two-piece colors seems remarkably effective and interesting for three-person play. When playing Triskelion, it is common to reach quite interesting endgames, full of tension and desperate tactics, that produce the winner only when the board is almost completely full.

Here are the initial moves of a game:

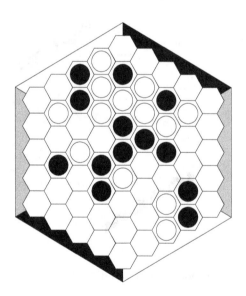

In this position, Grey has two possible paths to hope realistically to win (an upper path with white stones and a middle path with black stones). White can only win with black stones and shares some of that path with Grey. Eventually he may play a white stone at g7 hurting White and helping Black. However, in the next turn, Black must play a black stone, which is useless to him but may be helpful to his adversaries, giving some time to the others to prevent his vertical white path. This means Grey can profit from the adversaries' efforts.

Most played pieces are possible paths to the other players as well. Due to the possibility of making paths of both colors, it is rare for a player to be in a totally lost position. Using PDP is only possible very near the endgame and is limited by the fact that each player must play a specific color on the board. So, a player may create a position where the opponents cannot attack him because he may win the game by using either color, and, so, by trying to prevent him, they are indeed helping him to conclude the victorious connection.

Iqishiqi for Three

Another game with shared pieces is the three-player version of Iqishiqi (see p. 89.)

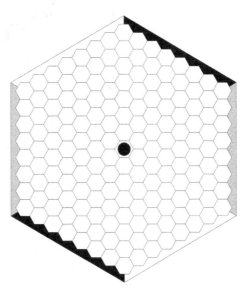

From the original rules, we modify the goal:

1. MOVER wins if NEXT cannot make a legal move on his turn.

2. If the stone reaches a corner owned by MOVER, he wins; otherwise, NEXT wins.

3. If the stone reaches an edge, the owner of that edge wins.

In this game, the STOP-NEXT rule applies only to reaching an edge, not to the ability to make a stalemating move.

Notes

This game is like building an elaborate maze of shared stones and trying to move the neutral stone into a dead end or making the adversaries push the neutral stone near one of your edges, where you might score a "goal" by landing it right on the edge.

Here is a sample game. This is the board at the eigth turn, (Black started, followed by Grey). The neutral stone started where piece 2 is. Move 6 sent the neutral to the south. Move 17 sent the neutral stone five hexagons to northwest. Move 18 sent the neutral five hexagons to the north. At the end of turn 6, the ball was on the bottom left sector but quickly changed to a totally different area (after moves 17 and 18). This happens usually when the local density of pieces increase to a certain threshold.

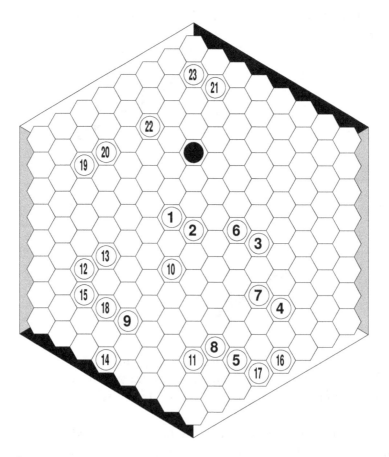

After three more turns:

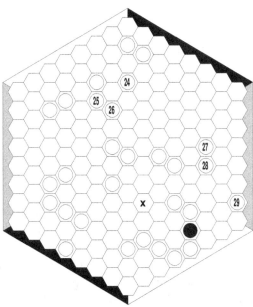

Now White can win by playing a stone at hexagon X, moving the ball to his edge. Note that Grey's move 29 was forced. Otherwise, it would have been illegal by STOP-NEXT.

Unequal Tasks

One of the intriguing things about three-player games is that it is not necessarily a bad thing if the players have somewhat different tasks, perhaps of differing difficulty. Though this would usually be fatal in a two-player game, the Tall Poppy effect ensures that with three, the easier tasks will not automatically lead to easier (or any) wins.

An example is the following game, adapted by Bill Taylor from one of his two-player inventions. Here the players have tasks differing noticeably in difficulty, but not so much that it might make it particularly easy for any of them to maintain the upper hand for long.

Porus Torus

This is a connection game played on a rectangular array of hexagons, representing a torus, with normal local Hex-connectivity. The top and bottom rows are adjacent, as the leftmost and rightmost columns. As in Triskelion, the three players play in fixed order, and alternately with black and white stones, with a view to making, in either color, their own sort of winning loop.

169

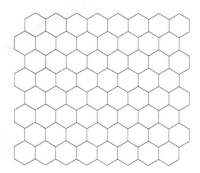

It is a fascinating fact that, on a torus, there are three kinds of "global" loop that can be made, that is, a loop going 'right around" the torus and not smoothly contractible to a point. Even better, the three types are mutually exclusive and exhaustive, meaning that one of the three must eventually be made, and it will not be possible to make two in different colors. This is a key property of good connection games (as seen in Hex). The three types of global loop, from easiest to hardest, are vertical, horizontal and spiral (check the next diagrams.)

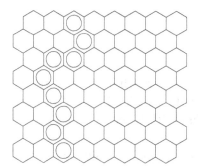 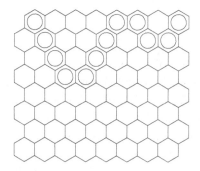

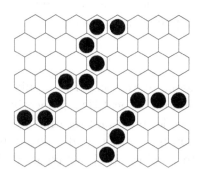

It is still possible that loops of two types in the same color could be made with a single move. In such cases, the rule is MOVER wins if he is one of the loop types; otherwise NEXT.

Rules

Players Diagonal, Horizontal, Vertical play in that fixed move order, dropping a stone on an empty hexagon. The stone color must be different from the stone dropped by the previous player. The first dropped stone is black.

Goals

If a global loop is formed in a single color, then the winner is Diagonal, Horizontal or Vertical, for a diagonal (spiral), horizontal, vertical loop respectively.

If two types of loop are made simultaneously, then MOVER wins if one of the types was his, otherwise NEXT wins.

The STOP-NEXT rule applies.

Notes

In the following sample game Diagonal wins in spite of having the last move and the hardest task:

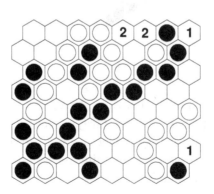

Diagonal wins because a group of black stones almost completed exists (missing only the hexagons marked 1). A group of white stones also almost complete (hexagons 2) is also present. The other players, even acting in concert, cannot stop all those connections.

Old Games in New Boards

There are several failed attempts to extend Hex to a hexagonal board. The common problem is how to prevent deadlocks that stalemate the outcome. In this variant, when a deadlock is reached, one player is out. When two players are out, the remaining player wins.

Hex for Three

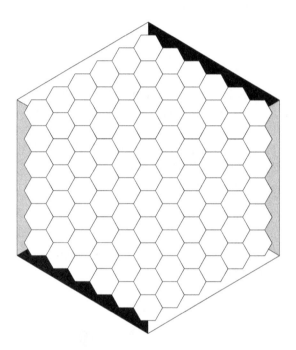

Rules

The game is played with Hex rules, except:

There are three players (White, Grey, Black) that, on each turn, drop a stone of his color on an empty hexagon. As soon as it becomes impossible for a player to ever complete any winning path of his own, he loses and stops moving, but his pieces remain.

Goal

A player wins if he creates a connection of stones of his color between his two edges, or if he eliminates all of the adversaries.

Notes

In the next diagram, we can check a game won by White. (The first to play was Grey, then White, then Black):

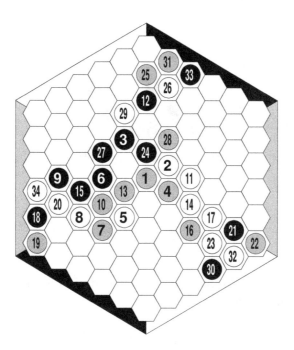

Move 33 eliminates Grey, which now is incapable to connect his edges. Move 34, by White, eliminates Black, and so White wins. Black came in second, and Grey is third.

This game is a mix of connection and blockage game. Players tend to concentrate on leaving as many as possible open paths to victory, instead of racing to connect (which would cause a sequence of blocking replies from the others, and one player cannot force a way through two others in concert). The enjoyable twist of this 3-Hex is the use of deadlocks as an essential part of the game dynamics. In an Aikido act, the problem became the solution.

Reversi for Three

Reversi is a quite successful commercial game. The rules are simple:

1. Each player drops one stone of his color in an empty square such that there are captures. (Check next rule.) If no captures are possible, the player must pass.

2. All enemy pieces between the dropped stone and another friendly stone (in any horizontal, vertical or diagonal direction) are captured and replaced by friendly stones.

In the next diagram, black stone 1 was dropped, producing the captures of all marked white stones.

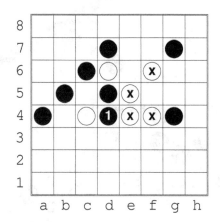

Reversi for three can be played very successfully on a hexagonal board. We suggest the initial setup:

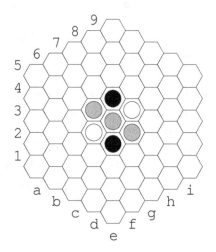

There is no STOP-NEXT rule, but we find it advisable to play with the extra rule so that no one may completely eliminate either opponent (unless this is unavoidable); this prevents two players deliberately or inadvertently ganging up on the other, to eliminate him early on. A player with no legal move merely passes, as in normal Reversi.

Balancing Turns

Sometimes players in three-player games may feel annoyance at the fixed order of moves, in that they may feel that one is a bad player. So, it is an advantage to be following that player in the move order. One possible solution is to allow MOVER to decide who plays NEXT. However, this could have the perverse effect of letting two players dictate that each other moves next, effectively starving the other one.

One remedy is to let players grow in strength with each turn that they are neither the nominator nor the nominee. For instance, if A nominates B to move next, then C gets to place an extra piece on the board. This strikes a balance between the benefits of moving (mobility) and the benefits of not moving (additional resources).

The nomination should not be completely arbitrary, of course, but should be dependent on the kind of move made. With these ideas in mind, Triad (by Cameron Browne) was born.

Triad

The game is played on the following three color hexagonal board:

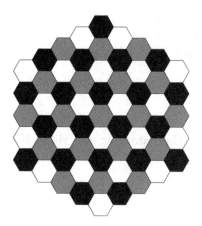

Here is the following initial setup:

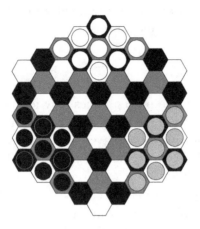

Rules

On each turn MOVER, the current player, must move, capture and drop. The current player must move one of their pieces in a straight line in any of the six hexagonal directions, to land on an empty foreign cell; any intervening cells must also be empty. The opponent who owns the landing cell becomes the Candidate, and the other opponent becomes the Bunny.

All opponents' pieces immediately adjacent to the landing cell are captured and removed from the board. The current player must make the move that captures the most pieces during each turn, but may choose amongst equals. This is called the max-capture rule.

MOVER must then drop a Bunny piece into any empty cell unless this player has just been eliminated.

The Candidate becomes the NEXT player to move.

Goal

Play stops the moment when any player is eliminated. The player with the most pieces left wins the game, or else there is a tie between the two remaining players if they are left with the same number of pieces.

Notes

One of the intriguing aspects of the game is the way in which micro-alliances form and evaporate between competing players with each passing turn, due to the Tall Poppy effect. As soon as one player edges ahead, the other two tend to cooperate briefly to pull the leader back until one of these two gains the upper hand and thus becomes the focus of attention.

The max-capture rule allows an interesting twist on this theme; clever players can manipulate their opponents into returning control of the move next turn or even can eliminate each other. This can be achieved through judicious placement of the Bunny piece. The next diagram shows an example:

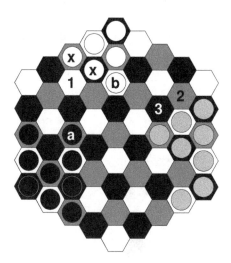

It is Black's turn. In this position, he is forced to capture two stones. (There are several options.) Black chooses to move piece A to hexagon 1, capturing two white stones (marked with an X). Since hexagon 1 is white, it will be White's turn. To end his turn, Black drops a grey stone in hexagon 2. So, Black, with this move, forced White to capture three stones by moving piece B to hexagon 3, giving the turn back to Black.

Simple Extensions

Gomoku for Three

This version is played on a 19×19 board, where each player, on each turn, drops a stone of his color on an empty intersection. The first player to make a row of four or more adjacent stones of his own color, orthogonally or diagonally, wins.

Notes

The key point of interest in this game is the STOP-NEXT rule, without which the game is useless, but which has both enabling and restrictive good effects when used. A hindering effect is that it becomes very dangerous for a player to make a single one-ended three. NEXT will ignore it and make his own open three if he can, forcing OTHER to block the original MOVER's three. This leaves MOVER to try to block both ends of NEXT's open three.

The next diagram shows an example of this. Black, after stone 1, makes White the winner. How? White can easily play at 2 since it is Grey that must stop Black. So, White creates an open line of three white stones that Black cannot stop on his next turn (if he plays at 4, White wins at 5).

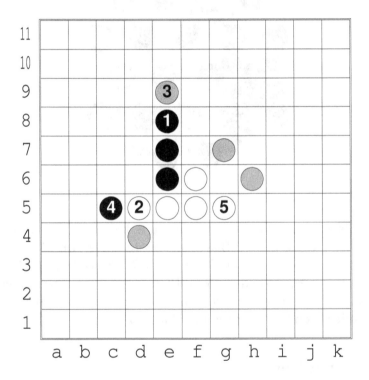

The helpful effect is that a player can often use the rule to keep his NEXT opponent busy, and maybe even both opponents, leaving time for his own constructive work.

Here is a sample showing the aggressive and fast nature of this game:

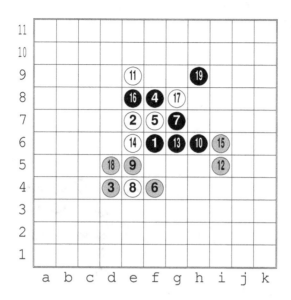

After a forced sequence (since stone 10), White must play at 20 to prevent a double open three that would give the game to Grey. Then:

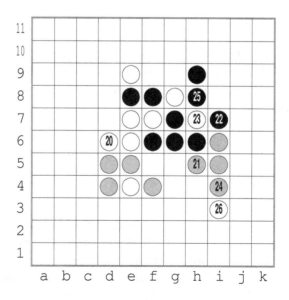

In this position, Grey must play 27 because of STOP-NEXT. (He cannot give an immediate win to Black.)

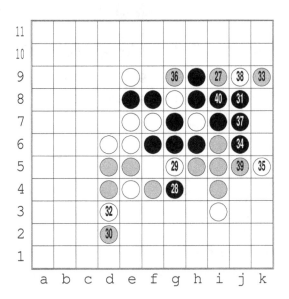

In this last diagram, move 30 forces White to play at 32, allowing Black's 31 and winning the game after a sequence of forced moves, so that in move 40 it manages a triple threat (g10, k6 and k8) which cannot be stopped by the other players.

Gonnect for Three

Gonnect (see p. 67) for three players provide connection and spatial concerns take their places in the multiplayer scenario. The rules are identical to the original game. All stones without any liberty are captured. This means a player may capture, in a single move, pieces from both adversaries. In the next diagram, if White moves at g9, he will capture the black and the grey stones.

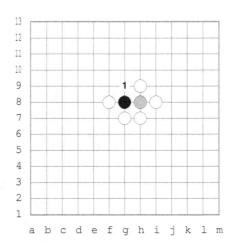

To inhibit PDP, there is an extra rule called *mandatory prioritized capture* which states that it is mandatory to capture NEXT stones (if any are available), and, if not, then the player must capture OTHER stones if possible. Providing these conditions are not met, he may choose freely. The rule states also that ALL opponents' liberty-less stones are captured, and so, it may be possible to capture both colors at once.

This provides some control over the next moves of both opponents, especially NEXT, which sometimes translates into some nice positions with several forced-group sequences.

Here is a position of a game that ended in a remarkable sequence of forced moves. Grey's turn is before White's.

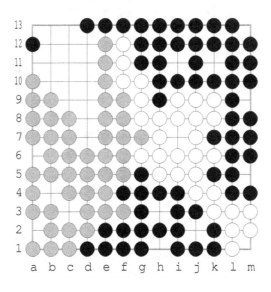

In this position, Black can still win with a horizontal connection while Grey can also win horizontally. White is in a very bad shape. Grey continues at c12:

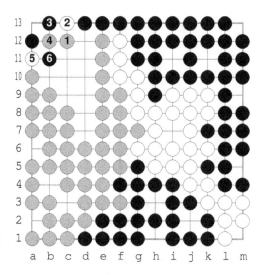

Piece 3 captures 2 and piece 6 captures 5. Moves 2 and 5 forced the next player to capture (namely, white stones). Then all subsequent moves produced forced captures!

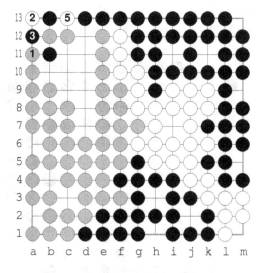

Piece 2 captures a12, piece 3 captures 2 (the same intersection where piece 4 was dropped and captured piece 3). Piece 5 captures b13. After this set of moves, we get the next board position:

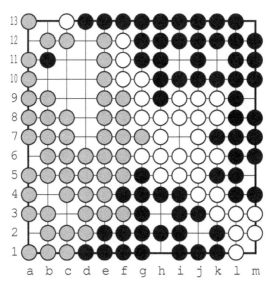

Now, Grey is already too strong.

Multi-Player Extensions

These games, and many others can be extended to four or more players with some success. Most commercial games are specifically designed for groups of four people or more. We believe that abstract games like these can have a place among a group of people after dinner, or on a rainy day.

Some games, of course, are very obviously designed specifically for three players or are virtually impossible to extend further. Triskelion and Triad are of this type, as is Hex. Some seem to have no hindrance at all in extending almost without limit, providing only that the board be big enough. Such types are Gomoku, Gonnect, and Reversi. Then there are some that may extend a little only. Porous Torus can extend to 4 players and Iqishiqi to 4, 5, or 6. However, we leave it to a further generation of gamers to investigate these matters more fully.

Glossary

Adjacency — Two pieces are adjacent if they occupy cells next to each other. The most common adjacencies in square boards are orthogonal and diagonal. In hexagonal boards, the concept is more natural since there is no such distinction.

Army — the player's friendly pieces.

Capture — the confiscation of one or more opponent pieces. The rules for capturing are game-dependent. The notation to indicate a capture usually consists of the starting cell and the final cell description, separated by a ":". So, a2:b3 means that the piece at a2 captured the piece at b3.

Cell — a board point, such as a square, a hexagon or an intersection.

Colors — In the tradition of chess and Go, the chosen colors for two players are White and Black. When necessary, the third color is Grey.

Column — a vertical line of cells in a board.

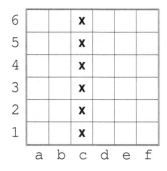

Connection — Two pieces are connected according to criteria specified by the game rules. The most common criterion is adjacency: two adjacent pieces are connected. This property is usually transitive, i.e., if piece A is connected to piece B, and B is connected to C, then A is connected to C.

Corners — specific board cells. The following diagrams show the corners of a square and a hexagonal board.

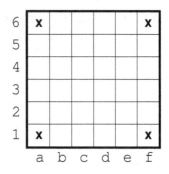

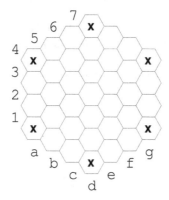

Coordinates — a syntactical way to identify a board cell. This book uses two different types of coordinates, one for square boards and another for hexagonal boards. For square boards, each cell is identified by the row (a number) and column (a letter), as in chess. For hexagonal boards, each cell is identified by a column (a letter) and by the descendent diagonal (a number). In the following diagrams, cell c5 is marked.

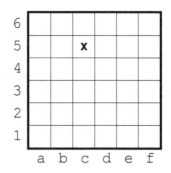

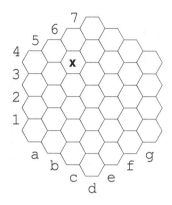

Custodian, Capture by — When a piece moves to a cell, forming a line of pieces, with friendly stones in the extremities and enemy stones in the middle, these enemy pieces are captured. In the next diagram, the black stone at e4, by moving to e3, may capture two white stones by a custodian capture.

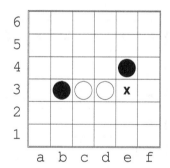

Dagger — A dagger represents the capacity to execute two consecutive moves. A dagger is not allowed to be used in the winning turn. Since it is a very powerful mechanism, it is normal to include some restrictions (e.g., it is not valid to use the dagger in two consecutive turns).

Diagonal — In square boards, the diagonal cells are those that touch the corners, but not the sides.

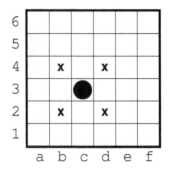

Edges — sets of cells not adjacent to other cells in all possible directions. The next diagrams show the edges of a square and hexagonal board.

	a	b	c	d	e	f
6	x	x	x	x	x	x
5	x					x
4	x					x
3	x					x
2	x					x
1	x	x	x	x	x	x

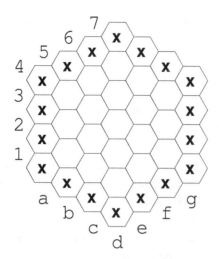

Empty — A cell is empty if it is not occupied by friendly, enemy, or neutral pieces.

Endgame — the final part of a game, its last moves. Also the last board position.

Enemy piece — an adversary's piece.

Freedom — a group is free when adjacent to at least one non-occupied cell.

Friendly piece — a player's piece.

Group — a connected set of friendly pieces.

In a row — A line of connected stones. The next diagram shows a four-in-a-row.

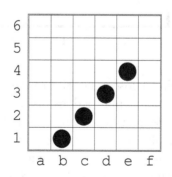

Isolate — A piece is isolated when it is not adjacent to any friendly piece.

Ko — a rule that forbids the repetition of the last (or before the last) board position. There is also a variant (called super-Ko), which forbids the repetition of every previous board position.

Liberty — A group has liberty if at least one of its pieces is adjacent to an empty cell.

n-tuple — See Tuple.

Neutral piece — a piece that is neither a friendly piece nor an enemy piece.

Orthogonal — horizontal or vertical.

Pass — A pass happens when a player skips his turn. Usually a game ends when both players consecutively pass.

Pie rule — a protocol to achieve a balanced game start. Usually, this rule is stated like: one player makes N moves (with both pieces), and the other player decides his own color based on those moves.

Precedence — To say a move of type A takes precedence over a move of type B is to say that whenever A is possible, it is illegal to play B.

Prisoners — captured enemy stones.

Row — a horizontal line of cells in a square board.

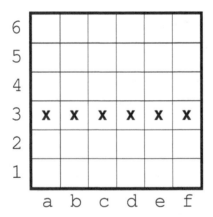

Sacrifice — when a player loses material to gain positional advantage.

Stalemate — when the next player does not have any valid move.

Stalemate, Win by — to win by stalemating the adversary.

Strategy — a guiding plan to achieve the player's goals.

Substitution, Capture by — when a piece moves to a cell occupied by an enemy piece, and the enemy piece is removed from play. The most famous game with this type of capture is chess.

Suicide — when a friendly piece is captured by the player's own move.

Tactic — a typical move to gain immediate material or positional advantage.

Territory — a set of cells that belong to one player according to given criteria. Usually, it is based on the siege of those cells by friendly pieces (and the edges). See the game of Go (p. 51) for the traditional definition of territory.

Tuple — (or n-tuple) an ordered set of n numbers. A tuple is represented by $(x_1, \ldots, x_n)$ where $x_1, \ldots, x_n$ are numbers. When $n = 2$, a tuple is called a ordered pair.

Turn — a pair of valid moves, one from the first player and another from the second player. Playing a game consists of executing consecutive turns.

Tweedledum and Tweedledee — tactical movement of imitating the last adversary move (by symmetry or by other means).

Y — a group of pieces connecting three edges.

Bibliography

[BEL]   BELL, R. C., *Board and Table Games from Many Civilizations*, Dover, 1979.

[BCG]   BERLEKAMP, E. R., CONWAY, J. H., GUY, R. K., *Winning Ways*, A. K. Peters, 2001.

[BIN]   BINMORE, K., *Fun and Games*, D. C. Heath and Co., 1992.

[BOU]   BOUTON, "Nim, a Game with Complete Mathematical Theory", *Annals of Mathematics*, Princeton (2) **3**, 1901/1902: 35–39.

[BRO]   BROWNE, C., *Connection games*, A. K. Peters, 2004

[CON]   CONWAY, J. H., *On Numbers and Games*, A. K. Peters, 2001.

[FAL]   FALKENER, E., *Games Ancient and Oriental*, Dover, 1961.

[MUR]   MURRAY, H. J. R., *A History of Board-Games other than Chess*, Oxford University Press, 1951.

[PAR]   PARLETT, D., *The Oxford History of Board Games*, Oxford University Press, 1999.

[SAC]   SACKSON, S., *A Gamut of Games*, Random House, 1969.

[JNS]   SILVA, JORGE NUNO, "Jogórios," *Boletim da Sociedade Portuguesa de Matemática*, 40, May 1999: 57–68.

[STR]   STRAFFIN, P.D., "Three Person Winner-Take-All Games with McCarthy's Revenge Rule," *College Math. Journal*, 16:5, 1985.

[THO]   THOMPSON, M., "Defining the abstract,"
[www.thegamesjournal.com/articles/DefiningtheAbstract.shtml], 2000.

A CATALOG OF SELECTED DOVER
BOOKS IN ALL FIELDS OF INTEREST

100 BEST-LOVED POEMS, Edited by Philip Smith. "The Passionate Shepherd to His Love," "Shall I compare thee to a summer's day?" "Death, be not proud," "The Raven," "The Road Not Taken," plus works by Blake, Wordsworth, Byron, Shelley, Keats, many others. 96pp. 5³⁄₁₆ x 8¼. 0-486-28553-7

100 SMALL HOUSES OF THE THIRTIES, Brown-Blodgett Company. Exterior photographs and floor plans for 100 charming structures. Illustrations of models accompanied by descriptions of interiors, color schemes, closet space, and other amenities. 200 illustrations. 112pp. 8⅜ x 11. 0-486-44131-8

1000 TURN-OF-THE-CENTURY HOUSES: With Illustrations and Floor Plans, Herbert C. Chivers. Reproduced from a rare edition, this showcase of homes ranges from cottages and bungalows to sprawling mansions. Each house is meticulously illustrated and accompanied by complete floor plans. 256pp. 9⅜ x 12¼.
0-486-45596-3

101 GREAT AMERICAN POEMS, Edited by The American Poetry & Literacy Project. Rich treasury of verse from the 19th and 20th centuries includes works by Edgar Allan Poe, Robert Frost, Walt Whitman, Langston Hughes, Emily Dickinson, T. S. Eliot, other notables. 96pp. 5³⁄₁₆ x 8¼. 0-486-40158-8

101 GREAT SAMURAI PRINTS, Utagawa Kuniyoshi. Kuniyoshi was a master of the warrior woodblock print — and these 18th-century illustrations represent the pinnacle of his craft. Full-color portraits of renowned Japanese samurais pulse with movement, passion, and remarkably fine detail. 112pp. 8⅜ x 11. 0-486-46523-3

ABC OF BALLET, Janet Grosser. Clearly worded, abundantly illustrated little guide defines basic ballet-related terms: arabesque, battement, pas de chat, relevé, sissonne, many others. Pronunciation guide included. Excellent primer. 48pp. 4³⁄₁₆ x 5¾.
0-486-40871-X

ACCESSORIES OF DRESS: An Illustrated Encyclopedia, Katherine Lester and Bess Viola Oerke. Illustrations of hats, veils, wigs, cravats, shawls, shoes, gloves, and other accessories enhance an engaging commentary that reveals the humor and charm of the many-sided story of accessorized apparel. 644 figures and 59 plates. 608pp. 6⅛ x 9¼.
0-486-43378-1

ADVENTURES OF HUCKLEBERRY FINN, Mark Twain. Join Huck and Jim as their boyhood adventures along the Mississippi River lead them into a world of excitement, danger, and self-discovery. Humorous narrative, lyrical descriptions of the Mississippi valley, and memorable characters. 224pp. 5³⁄₁₆ x 8¼. 0-486-28061-6

ALICE STARMORE'S BOOK OF FAIR ISLE KNITTING, Alice Starmore. A noted designer from the region of Scotland's Fair Isle explores the history and techniques of this distinctive, stranded-color knitting style and provides copious illustrated instructions for 14 original knitwear designs. 208pp. 8⅜ x 10⅞. 0-486-47218-3

Browse over 9,000 books at www.doverpublications.com

THE JUNGLE, Upton Sinclair. 1906 bestseller shockingly reveals intolerable labor practices and working conditions in the Chicago stockyards as it tells the grim story of a Slavic family that emigrates to America full of optimism but soon faces despair. 320pp. 5³⁄₁₆ x 8¼. 0-486-41923-1

THE KINGDOM OF GOD IS WITHIN YOU, Leo Tolstoy. The soul-searching book that inspired Gandhi to embrace the concept of passive resistance, Tolstoy's 1894 polemic clearly outlines a radical, well-reasoned revision of traditional Christian thinking. 352pp. 5³⁄₁₆ x 8¼. 0-486-45138-0

THE LADY OR THE TIGER?: and Other Logic Puzzles, Raymond M. Smullyan. Created by a renowned puzzle master, these whimsically themed challenges involve paradoxes about probability, time, and change; metapuzzles; and self-referentiality. Nineteen chapters advance in difficulty from relatively simple to highly complex. 1982 edition. 240pp. 5⅜ x 8½. 0-486-47027-X

LEAVES OF GRASS: The Original 1855 Edition, Walt Whitman. Whitman's immortal collection includes some of the greatest poems of modern times, including his masterpiece, "Song of Myself." Shattering standard conventions, it stands as an unabashed celebration of body and nature. 128pp. 5³⁄₁₆ x 8¼. 0-486-45676-5

LES MISÉRABLES, Victor Hugo. Translated by Charles E. Wilbour. Abridged by James K. Robinson. A convict's heroic struggle for justice and redemption plays out against a fiery backdrop of the Napoleonic wars. This edition features the excellent original translation and a sensitive abridgment. 304pp. 6⅛ x 9¼.
0-486-45789-3

LILITH: A Romance, George MacDonald. In this novel by the father of fantasy literature, a man travels through time to meet Adam and Eve and to explore humanity's fall from grace and ultimate redemption. 240pp. 5⅜ x 8½.
0-486-46818-6

THE LOST LANGUAGE OF SYMBOLISM, Harold Bayley. This remarkable book reveals the hidden meaning behind familiar images and words, from the origins of Santa Claus to the fleur-de-lys, drawing from mythology, folklore, religious texts, and fairy tales. 1,418 illustrations. 784pp. 5⅜ x 8½. 0-486-44787-1

MACBETH, William Shakespeare. A Scottish nobleman murders the king in order to succeed to the throne. Tortured by his conscience and fearful of discovery, he becomes tangled in a web of treachery and deceit that ultimately spells his doom. 96pp. 5³⁄₁₆ x 8¼. 0-486-27802-6

MAKING AUTHENTIC CRAFTSMAN FURNITURE: Instructions and Plans for 62 Projects, Gustav Stickley. Make authentic reproductions of handsome, functional, durable furniture: tables, chairs, wall cabinets, desks, a hall tree, and more. Construction plans with drawings, schematics, dimensions, and lumber specs reprinted from 1900s *The Craftsman* magazine. 128pp. 8⅛ x 11. 0-486-25000-8

MATHEMATICS FOR THE NONMATHEMATICIAN, Morris Kline. Erudite and entertaining overview follows development of mathematics from ancient Greeks to present. Topics include logic and mathematics, the fundamental concept, differential calculus, probability theory, much more. Exercises and problems. 641pp. 5⅜ x 8½. 0-486-24823-2

MEMOIRS OF AN ARABIAN PRINCESS FROM ZANZIBAR, Emily Ruete. This 19th-century autobiography offers a rare inside look at the society surrounding a sultan's palace. A real-life princess in exile recalls her vanished world of harems, slave trading, and court intrigues. 288pp. 5⅜ x 8½. 0-486-47121-7

Browse over 9,000 books at www.doverpublications.com

THE METAMORPHOSIS AND OTHER STORIES, Franz Kafka. Excellent new English translations of title story (considered by many critics Kafka's most perfect work), plus "The Judgment," "In the Penal Colony," "A Country Doctor," and "A Report to an Academy." Note. 96pp. 5³⁄₁₆ x 8¼. 0-486-29030-1

MICROSCOPIC ART FORMS FROM THE PLANT WORLD, R. Anheisser. From undulating curves to complex geometrics, a world of fascinating images abound in this classic, illustrated survey of microscopic plants. Features 400 detailed illustrations of nature's minute but magnificent handiwork. The accompanying CD-ROM includes all of the images in the book. 128pp. 9 x 9. 0-486-46013-4

A MIDSUMMER NIGHT'S DREAM, William Shakespeare. Among the most popular of Shakespeare's comedies, this enchanting play humorously celebrates the vagaries of love as it focuses upon the intertwined romances of several pairs of lovers. Explanatory footnotes. 80pp. 5³⁄₁₆ x 8¼. 0-486-27067-X

THE MONEY CHANGERS, Upton Sinclair. Originally published in 1908, this cautionary novel from the author of *The Jungle* explores corruption within the American system as a group of power brokers joins forces for personal gain, triggering a crash on Wall Street. 192pp. 5⅜ x 8½. 0-486-46917-4

THE MOST POPULAR HOMES OF THE TWENTIES, William A. Radford. With a New Introduction by Daniel D. Reiff. Based on a rare 1925 catalog, this architectural showcase features floor plans, construction details, and photos of 26 homes, plus articles on entrances, porches, garages, and more. 250 illustrations, 21 color plates. 176pp. 8⅜ x 11. 0-486-47028-8

MY 66 YEARS IN THE BIG LEAGUES, Connie Mack. With a New Introduction by Rich Westcott. A Founding Father of modern baseball, Mack holds the record for most wins — and losses — by a major league manager. Enhanced by 70 photographs, his warmhearted autobiography is populated by many legends of the game. 288pp. 5⅜ x 8½. 0-486-47184-5

NARRATIVE OF THE LIFE OF FREDERICK DOUGLASS, Frederick Douglass. Douglass's graphic depictions of slavery, harrowing escape to freedom, and life as a newspaper editor, eloquent orator, and impassioned abolitionist. 96pp. 5³⁄₁₆ x 8¼. 0-486-28499-9

THE NIGHTLESS CITY: Geisha and Courtesan Life in Old Tokyo, J. E. de Becker. This unsurpassed study from 100 years ago ventured into Tokyo's red-light district to survey geisha and courtesan life and offer meticulous descriptions of training, dress, social hierarchy, and erotic practices. 49 black-and-white illustrations; 2 maps. 496pp. 5⅜ x 8½. 0-486-45563-7

THE ODYSSEY, Homer. Excellent prose translation of ancient epic recounts adventures of the homeward-bound Odysseus. Fantastic cast of gods, giants, cannibals, sirens, other supernatural creatures — true classic of Western literature. 256pp. 5³⁄₁₆ x 8¼. 0-486-40654-7

OEDIPUS REX, Sophocles. Landmark of Western drama concerns the catastrophe that ensues when King Oedipus discovers he has inadvertently killed his father and married his mother. Masterly construction, dramatic irony. Explanatory footnotes. 64pp. 5³⁄₁₆ x 8¼. 0-486-26877-2

ONCE UPON A TIME: The Way America Was, Eric Sloane. Nostalgic text and drawings brim with gentle philosophies and descriptions of how we used to live — self-sufficiently — on the land, in homes, and among the things built by hand. 44 line illustrations. 64pp. 8⅜ x 11. 0-486-44411-2

ALICE'S ADVENTURES IN WONDERLAND, Lewis Carroll. Beloved classic about a little girl lost in a topsy-turvy land and her encounters with the White Rabbit, March Hare, Mad Hatter, Cheshire Cat, and other delightfully improbable characters. 42 illustrations by Sir John Tenniel. 96pp. 5³⁄₁₆ x 8¼. 0-486-27543-4

AMERICA'S LIGHTHOUSES: An Illustrated History, Francis Ross Holland. Profusely illustrated fact-filled survey of American lighthouses since 1716. Over 200 stations — East, Gulf, and West coasts, Great Lakes, Hawaii, Alaska, Puerto Rico, the Virgin Islands, and the Mississippi and St. Lawrence Rivers. 240pp. 8 x 10¾.
0-486-25576-X

AN ENCYCLOPEDIA OF THE VIOLIN, Alberto Bachmann. Translated by Frederick H. Martens. Introduction by Eugene Ysaye. First published in 1925, this renowned reference remains unsurpassed as a source of essential information, from construction and evolution to repertoire and technique. Includes a glossary and 73 illustrations. 496pp. 6⅛ x 9¼. 0-486-46618-3

ANIMALS: 1,419 Copyright-Free Illustrations of Mammals, Birds, Fish, Insects, etc., Selected by Jim Harter. Selected for its visual impact and ease of use, this outstanding collection of wood engravings presents over 1,000 species of animals in extremely lifelike poses. Includes mammals, birds, reptiles, amphibians, fish, insects, and other invertebrates. 284pp. 9 x 12. 0-486-23766-4

THE ANNALS, Tacitus. Translated by Alfred John Church and William Jackson Brodribb. This vital chronicle of Imperial Rome, written by the era's great historian, spans A.D. 14-68 and paints incisive psychological portraits of major figures, from Tiberius to Nero. 416pp. 5³⁄₁₆ x 8¼. 0-486-45236-0

ANTIGONE, Sophocles. Filled with passionate speeches and sensitive probing of moral and philosophical issues, this powerful and often-performed Greek drama reveals the grim fate that befalls the children of Oedipus. Footnotes. 64pp. 5³⁄₁₆ x 8 ¼. 0-486-27804-2

ART DECO DECORATIVE PATTERNS IN FULL COLOR, Christian Stoll. Reprinted from a rare 1910 portfolio, 160 sensuous and exotic images depict a breathtaking array of florals, geometrics, and abstracts — all elegant in their stark simplicity. 64pp. 8⅜ x 11. 0-486-44862-2

THE ARTHUR RACKHAM TREASURY: 86 Full-Color Illustrations, Arthur Rackham. Selected and Edited by Jeff A. Menges. A stunning treasury of 86 full-page plates span the famed English artist's career, from *Rip Van Winkle* (1905) to masterworks such as *Undine, A Midsummer Night's Dream,* and *Wind in the Willows* (1939). 96pp. 8⅜ x 11.
0-486-44685-9

THE AUTHENTIC GILBERT & SULLIVAN SONGBOOK, W. S. Gilbert and A. S. Sullivan. The most comprehensive collection available, this songbook includes selections from every one of Gilbert and Sullivan's light operas. Ninety-two numbers are presented uncut and unedited, and in their original keys. 410pp. 9 x 12.
0-486-23482-7

THE AWAKENING, Kate Chopin. First published in 1899, this controversial novel of a New Orleans wife's search for love outside a stifling marriage shocked readers. Today, it remains a first-rate narrative with superb characterization. New introductory Note. 128pp. 5³⁄₁₆ x 8¼. 0-486-27786-0

BASIC DRAWING, Louis Priscilla. Beginning with perspective, this commonsense manual progresses to the figure in movement, light and shade, anatomy, drapery, composition, trees and landscape, and outdoor sketching. Black-and-white illustrations throughout. 128pp. 8⅜ x 11. 0-486-45815-6

Browse over 9,000 books at www.doverpublications.com

THE BATTLES THAT CHANGED HISTORY, Fletcher Pratt. Historian profiles 16 crucial conflicts, ancient to modern, that changed the course of Western civilization. Gripping accounts of battles led by Alexander the Great, Joan of Arc, Ulysses S. Grant, other commanders. 27 maps. 352pp. 5⅜ x 8½.   0-486-41129-X

BEETHOVEN'S LETTERS, Ludwig van Beethoven. Edited by Dr. A. C. Kalischer. Features 457 letters to fellow musicians, friends, greats, patrons, and literary men. Reveals musical thoughts, quirks of personality, insights, and daily events. Includes 15 plates. 410pp. 5⅜ x 8½.   0-486-22769-3

BERNICE BOBS HER HAIR AND OTHER STORIES, F. Scott Fitzgerald. This brilliant anthology includes 6 of Fitzgerald's most popular stories: "The Diamond as Big as the Ritz," the title tale, "The Offshore Pirate," "The Ice Palace," "The Jelly Bean," and "May Day." 176pp. 5⅜ x 8½.   0-486-47049-0

BESLER'S BOOK OF FLOWERS AND PLANTS: 73 Full-Color Plates from Hortus Eystettensis, 1613, Basilius Besler. Here is a selection of magnificent plates from the *Hortus Eystettensis,* which vividly illustrated and identified the plants, flowers, and trees that thrived in the legendary German garden at Eichstätt. 80pp. 8⅜ x 11.   0-486-46005-3

THE BOOK OF KELLS, Edited by Blanche Cirker. Painstakingly reproduced from a rare facsimile edition, this volume contains full-page decorations, portraits, illustrations, plus a sampling of textual leaves with exquisite calligraphy and ornamentation. 32 full-color illustrations. 32pp. 9⅜ x 12¼.   0-486-24345-1

THE BOOK OF THE CROSSBOW: With an Additional Section on Catapults and Other Siege Engines, Ralph Payne-Gallwey. Fascinating study traces history and use of crossbow as military and sporting weapon, from Middle Ages to modern times. Also covers related weapons: balistas, catapults, Turkish bows, more. Over 240 illustrations. 400pp. 7¼ x 10⅛.   0-486-28720-3

THE BUNGALOW BOOK: Floor Plans and Photos of 112 Houses, 1910, Henry L. Wilson. Here are 112 of the most popular and economic blueprints of the early 20th century — plus an illustration or photograph of each completed house. A wonderful time capsule that still offers a wealth of valuable insights. 160pp. 8⅜ x 11.   0-486-45104-6

THE CALL OF THE WILD, Jack London. A classic novel of adventure, drawn from London's own experiences as a Klondike adventurer, relating the story of a heroic dog caught in the brutal life of the Alaska Gold Rush. Note. 64pp. 5³⁄₁₆ x 8¼.   0-486-26472-6

CANDIDE, Voltaire. Edited by Francois-Marie Arouet. One of the world's great satires since its first publication in 1759. Witty, caustic skewering of romance, science, philosophy, religion, government — nearly all human ideals and institutions. 112pp. 5³⁄₁₆ x 8¼.   0-486-26689-3

CELEBRATED IN THEIR TIME: Photographic Portraits from the George Grantham Bain Collection, Edited by Amy Pastan. With an Introduction by Michael Carlebach. Remarkable portrait gallery features 112 rare images of Albert Einstein, Charlie Chaplin, the Wright Brothers, Henry Ford, and other luminaries from the worlds of politics, art, entertainment, and industry. 128pp. 8⅜ x 11.   0-486-46754-6

CHARIOTS FOR APOLLO: The NASA History of Manned Lunar Spacecraft to 1969, Courtney G. Brooks, James M. Grimwood, and Loyd S. Swenson, Jr. This illustrated history by a trio of experts is the definitive reference on the Apollo spacecraft and lunar modules. It traces the vehicles' design, development, and operation in space. More than 100 photographs and illustrations. 576pp. 6¾ x 9¼. 0-486-46756-2

A CHRISTMAS CAROL, Charles Dickens. This engrossing tale relates Ebenezer Scrooge's ghostly journeys through Christmases past, present, and future and his ultimate transformation from a harsh and grasping old miser to a charitable and compassionate human being. 80pp. 5³⁄₁₆ x 8¼. 0-486-26865-9

COMMON SENSE, Thomas Paine. First published in January of 1776, this highly influential landmark document clearly and persuasively argued for American separation from Great Britain and paved the way for the Declaration of Independence. 64pp. 5³⁄₁₆ x 8¼. 0-486-29602-4

THE COMPLETE SHORT STORIES OF OSCAR WILDE, Oscar Wilde. Complete texts of "The Happy Prince and Other Tales," "A House of Pomegranates," "Lord Arthur Savile's Crime and Other Stories," "Poems in Prose," and "The Portrait of Mr. W. H." 208pp. 5³⁄₁₆ x 8¼. 0-486-45216-6

COMPLETE SONNETS, William Shakespeare. Over 150 exquisite poems deal with love, friendship, the tyranny of time, beauty's evanescence, death, and other themes in language of remarkable power, precision, and beauty. Glossary of archaic terms. 80pp. 5³⁄₁₆ x 8¼. 0-486-26686-9

THE COUNT OF MONTE CRISTO: Abridged Edition, Alexandre Dumas. Falsely accused of treason, Edmond Dantès is imprisoned in the bleak Chateau d'If. After a hair-raising escape, he launches an elaborate plot to extract a bitter revenge against those who betrayed him. 448pp. 5³⁄₁₆ x 8¼. 0-486-45643-9

CRAFTSMAN BUNGALOWS: Designs from the Pacific Northwest, Yoho & Merritt. This reprint of a rare catalog, showcasing the charming simplicity and cozy style of Craftsman bungalows, is filled with photos of completed homes, plus floor plans and estimated costs. An indispensable resource for architects, historians, and illustrators. 112pp. 10 x 7. 0-486-46875-5

CRAFTSMAN BUNGALOWS: 59 Homes from "The Craftsman," Edited by Gustav Stickley. Best and most attractive designs from Arts and Crafts Movement publication — 1903–1916 — includes sketches, photographs of homes, floor plans, descriptive text. 128pp. 8¼ x 11. 0-486-25829-7

CRIME AND PUNISHMENT, Fyodor Dostoyevsky. Translated by Constance Garnett. Supreme masterpiece tells the story of Raskolnikov, a student tormented by his own thoughts after he murders an old woman. Overwhelmed by guilt and terror, he confesses and goes to prison. 480pp. 5³⁄₁₆ x 8¼. 0-486-41587-2

THE DECLARATION OF INDEPENDENCE AND OTHER GREAT DOCUMENTS OF AMERICAN HISTORY: 1775-1865, Edited by John Grafton. Thirteen compelling and influential documents: Henry's "Give Me Liberty or Give Me Death," Declaration of Independence, The Constitution, Washington's First Inaugural Address, The Monroe Doctrine, The Emancipation Proclamation, Gettysburg Address, more. 64pp. 5³⁄₁₆ x 8¼. 0-486-41124-9

THE DESERT AND THE SOWN: Travels in Palestine and Syria, Gertrude Bell. "The female Lawrence of Arabia," Gertrude Bell wrote captivating, perceptive accounts of her travels in the Middle East. This intriguing narrative, accompanied by 160 photos, traces her 1905 sojourn in Lebanon, Syria, and Palestine. 368pp. 5⅜ x 8½. 0-486-46876-3

A DOLL'S HOUSE, Henrik Ibsen. Ibsen's best-known play displays his genius for realistic prose drama. An expression of women's rights, the play climaxes when the central character, Nora, rejects a smothering marriage and life in "a doll's house." 80pp. 5³⁄₁₆ x 8¼. 0-486-27062-9

Browse over 9,000 books at www.doverpublications.com

DOOMED SHIPS: Great Ocean Liner Disasters, William H. Miller, Jr. Nearly 200 photographs, many from private collections, highlight tales of some of the vessels whose pleasure cruises ended in catastrophe: the *Morro Castle, Normandie, Andrea Doria, Europa,* and many others. 128pp. 8⅞ x 11¾. 0-486-45366-9

THE DORÉ BIBLE ILLUSTRATIONS, Gustave Doré. Detailed plates from the Bible: the Creation scenes, Adam and Eve, horrifying visions of the Flood, the battle sequences with their monumental crowds, depictions of the life of Jesus, 241 plates in all. 241pp. 9 x 12. 0-486-23004-X

DRAWING DRAPERY FROM HEAD TO TOE, Cliff Young. Expert guidance on how to draw shirts, pants, skirts, gloves, hats, and coats on the human figure, including folds in relation to the body, pull and crush, action folds, creases, more. Over 200 drawings. 48pp. 8¼ x 11. 0-486-45591-2

DUBLINERS, James Joyce. A fine and accessible introduction to the work of one of the 20th century's most influential writers, this collection features 15 tales, including a masterpiece of the short-story genre, "The Dead." 160pp. 5³⁄₁₆ x 8¼. 0-486-26870-5

EASY-TO-MAKE POP-UPS, Joan Irvine. Illustrated by Barbara Reid. Dozens of wonderful ideas for three-dimensional paper fun — from holiday greeting cards with moving parts to a pop-up menagerie. Easy-to-follow, illustrated instructions for more than 30 projects. 299 black-and-white illustrations. 96pp. 8⅜ x 11. 0-486-44622-0

EASY-TO-MAKE STORYBOOK DOLLS: A "Novel" Approach to Cloth Dollmaking, Sherralyn St. Clair. Favorite fictional characters come alive in this unique beginner's dollmaking guide. Includes patterns for Pollyanna, Dorothy from *The Wonderful Wizard of Oz,* Mary of *The Secret Garden,* plus easy-to-follow instructions, 263 black-and-white illustrations, and an 8-page color insert. 112pp. 8¼ x 11. 0-486-47360-0

EINSTEIN'S ESSAYS IN SCIENCE, Albert Einstein. Speeches and essays in accessible, everyday language profile influential physicists such as Niels Bohr and Isaac Newton. They also explore areas of physics to which the author made major contributions. 128pp. 5 x 8. 0-486-47011-3

EL DORADO: Further Adventures of the Scarlet Pimpernel, Baroness Orczy. A popular sequel to *The Scarlet Pimpernel,* this suspenseful story recounts the Pimpernel's attempts to rescue the Dauphin from imprisonment during the French Revolution. An irresistible blend of intrigue, period detail, and vibrant characterizations. 352pp. 5³⁄₁₆ x 8¼. 0-486-44026-5

ELEGANT SMALL HOMES OF THE TWENTIES: 99 Designs from a Competition, Chicago Tribune. Nearly 100 designs for five- and six-room houses feature New England and Southern colonials, Normandy cottages, stately Italianate dwellings, and other fascinating snapshots of American domestic architecture of the 1920s. 112pp. 9 x 12. 0-486-46910-7

THE ELEMENTS OF STYLE: The Original Edition, William Strunk, Jr. This is the book that generations of writers have relied upon for timeless advice on grammar, diction, syntax, and other essentials. In concise terms, it identifies the principal requirements of proper style and common errors. 64pp. 5⅜ x 8½. 0-486-44798-7

THE ELUSIVE PIMPERNEL, Baroness Orczy. Robespierre's revolutionaries find their wicked schemes thwarted by the heroic Pimpernel — Sir Percival Blakeney. In this thrilling sequel, Chauvelin devises a plot to eliminate the Pimpernel and his wife. 272pp. 5³⁄₁₆ x 8¼. 0-486-45464-9

Browse over 9,000 books at www.doverpublications.com

AN ENCYCLOPEDIA OF BATTLES: Accounts of Over 1,560 Battles from 1479 B.C. to the Present, David Eggenberger. Essential details of every major battle in recorded history from the first battle of Megiddo in 1479 B.C. to Grenada in 1984. List of battle maps. 99 illustrations. 544pp. 6½ x 9¼. 0-486-24913-1

ENCYCLOPEDIA OF EMBROIDERY STITCHES, INCLUDING CREWEL, Marion Nichols. Precise explanations and instructions, clearly illustrated, on how to work chain, back, cross, knotted, woven stitches, and many more — 178 in all, including Cable Outline, Whipped Satin, and Eyelet Buttonhole. Over 1400 illustrations. 219pp. 8⅜ x 11¼. 0-486-22929-7

ENTER JEEVES: 15 Early Stories, P. G. Wodehouse. Splendid collection contains first 8 stories featuring Bertie Wooster, the deliciously dim aristocrat and Jeeves, his brainy, imperturbable manservant. Also, the complete Reggie Pepper (Bertie's prototype) series. 288pp. 5⅜ x 8½. 0-486-29717-9

ERIC SLOANE'S AMERICA: Paintings in Oil, Michael Wigley. With a Foreword by Mimi Sloane. Eric Sloane's evocative oils of America's landscape and material culture shimmer with immense historical and nostalgic appeal. This original hardcover collection gathers nearly a hundred of his finest paintings, with subjects ranging from New England to the American Southwest. 128pp. 10⅜ x 9.
0-486-46525-X

ETHAN FROME, Edith Wharton. Classic story of wasted lives, set against a bleak New England background. Superbly delineated characters in a hauntingly grim tale of thwarted love. Considered by many to be Wharton's masterpiece. 96pp. 5³⁄₁₆ x 8 ¼.
0-486-26690-7

THE EVERLASTING MAN, G. K. Chesterton. Chesterton's view of Christianity — as a blend of philosophy and mythology, satisfying intellect and spirit — applies to his brilliant book, which appeals to readers' heads as well as their hearts. 288pp. 5⅜ x 8½.
0-486-46036-3

THE FIELD AND FOREST HANDY BOOK, Daniel Beard. Written by a co-founder of the Boy Scouts, this appealing guide offers illustrated instructions for building kites, birdhouses, boats, igloos, and other fun projects, plus numerous helpful tips for campers. 448pp. 5³⁄₁₆ x 8¼. 0-486-46191-2

FINDING YOUR WAY WITHOUT MAP OR COMPASS, Harold Gatty. Useful, instructive manual shows would-be explorers, hikers, bikers, scouts, sailors, and survivalists how to find their way outdoors by observing animals, weather patterns, shifting sands, and other elements of nature. 288pp. 5⅜ x 8½. 0-486-40613-X

FIRST FRENCH READER: A Beginner's Dual-Language Book, Edited and Translated by Stanley Appelbaum. This anthology introduces 50 legendary writers — Voltaire, Balzac, Baudelaire, Proust, more — through passages from *The Red and the Black, Les Misérables, Madame Bovary,* and other classics. Original French text plus English translation on facing pages. 240pp. 5⅜ x 8½. 0-486-46178-5

FIRST GERMAN READER: A Beginner's Dual-Language Book, Edited by Harry Steinhauer. Specially chosen for their power to evoke German life and culture, these short, simple readings include poems, stories, essays, and anecdotes by Goethe, Hesse, Heine, Schiller, and others. 224pp. 5⅜ x 8½. 0-486-46179-3

FIRST SPANISH READER: A Beginner's Dual-Language Book, Angel Flores. Delightful stories, other material based on works of Don Juan Manuel, Luis Taboada, Ricardo Palma, other noted writers. Complete faithful English translations on facing pages. Exercises. 176pp. 5⅜ x 8½. 0-486-25810-6

Browse over 9,000 books at www.doverpublications.com

FIVE ACRES AND INDEPENDENCE, Maurice G. Kains. Great back-to-the-land classic explains basics of self-sufficient farming. The one book to get. 95 illustrations. 397pp. 5⅜ x 8½. 0-486-20974-1

FLAGG'S SMALL HOUSES: Their Economic Design and Construction, 1922, Ernest Flagg. Although most famous for his skyscrapers, Flagg was also a proponent of the well-designed single-family dwelling. His classic treatise features innovations that save space, materials, and cost. 526 illustrations. 160pp. 9⅜ x 12¼. 0-486-45197-6

FLATLAND: A Romance of Many Dimensions, Edwin A. Abbott. Classic of science (and mathematical) fiction — charmingly illustrated by the author — describes the adventures of A. Square, a resident of Flatland, in Spaceland (three dimensions), Lineland (one dimension), and Pointland (no dimensions). 96pp. 5³⁄₁₆ x 8¼. 0-486-27263-X

FRANKENSTEIN, Mary Shelley. The story of Victor Frankenstein's monstrous creation and the havoc it caused has enthralled generations of readers and inspired countless writers of horror and suspense. With the author's own 1831 introduction. 176pp. 5³⁄₁₆ x 8¼. 0-486-28211-2

THE GARGOYLE BOOK: 572 Examples from Gothic Architecture, Lester Burbank Bridaham. Dispelling the conventional wisdom that French Gothic architectural flourishes were born of despair or gloom, Bridaham reveals the whimsical nature of these creations and the ingenious artisans who made them. 572 illustrations. 224pp. 8⅜ x 11. 0-486-44754-5

THE GIFT OF THE MAGI AND OTHER SHORT STORIES, O. Henry. Sixteen captivating stories by one of America's most popular storytellers. Included are such classics as "The Gift of the Magi," "The Last Leaf," and "The Ransom of Red Chief." Publisher's Note. 96pp. 5³⁄₁₆ x 8¼. 0-486-27061-0

THE GOETHE TREASURY: Selected Prose and Poetry, Johann Wolfgang von Goethe. Edited, Selected, and with an Introduction by Thomas Mann. In addition to his lyric poetry, Goethe wrote travel sketches, autobiographical studies, essays, letters, and proverbs in rhyme and prose. This collection presents outstanding examples from each genre. 368pp. 5⅜ x 8½. 0-486-44780-4

GREAT EXPECTATIONS, Charles Dickens. Orphaned Pip is apprenticed to the dirty work of the forge but dreams of becoming a gentleman — and one day finds himself in possession of "great expectations." Dickens' finest novel. 400pp. 5³⁄₁₆ x 8¼. 0-486-41586-4

GREAT WRITERS ON THE ART OF FICTION: From Mark Twain to Joyce Carol Oates, Edited by James Daley. An indispensable source of advice and inspiration, this anthology features essays by Henry James, Kate Chopin, Willa Cather, Sinclair Lewis, Jack London, Raymond Chandler, Raymond Carver, Eudora Welty, and Kurt Vonnegut, Jr. 192pp. 5⅜ x 8½. 0-486-45128-3

HAMLET, William Shakespeare. The quintessential Shakespearean tragedy, whose highly charged confrontations and anguished soliloquies probe depths of human feeling rarely sounded in any art. Reprinted from an authoritative British edition complete with illuminating footnotes. 128pp. 5³⁄₁₆ x 8¼. 0-486-27278-8

THE HAUNTED HOUSE, Charles Dickens. A Yuletide gathering in an eerie country retreat provides the backdrop for Dickens and his friends — including Elizabeth Gaskell and Wilkie Collins — who take turns spinning supernatural yarns. 144pp. 5⅜ x 8½. 0-486-46309-5

Browse over 9,000 books at www.doverpublications.com

HEART OF DARKNESS, Joseph Conrad. Dark allegory of a journey up the Congo River and the narrator's encounter with the mysterious Mr. Kurtz. Masterly blend of adventure, character study, psychological penetration. For many, Conrad's finest, most enigmatic story. 80pp. 5³⁄₁₆ x 8¼. 0-486-26464-5

HENSON AT THE NORTH POLE, Matthew A. Henson. This thrilling memoir by the heroic African-American who was Peary's companion through two decades of Arctic exploration recounts a tale of danger, courage, and determination. "Fascinating and exciting." — *Commonweal.* 128pp. 5⅜ x 8½. 0-486-45472-X

HISTORIC COSTUMES AND HOW TO MAKE THEM, Mary Fernald and E. Shenton. Practical, informative guidebook shows how to create everything from short tunics worn by Saxon men in the fifth century to a lady's bustle dress of the late 1800s. 81 illustrations. 176pp. 5⅜ x 8½. 0-486-44906-8

THE HOUND OF THE BASKERVILLES, Arthur Conan Doyle. A deadly curse in the form of a legendary ferocious beast continues to claim its victims from the Baskerville family until Holmes and Watson intervene. Often called the best detective story ever written. 128pp. 5³⁄₁₆ x 8¼. 0-486-28214-7

THE HOUSE BEHIND THE CEDARS, Charles W. Chesnutt. Originally published in 1900, this groundbreaking novel by a distinguished African-American author recounts the drama of a brother and sister who "pass for white" during the dangerous days of Reconstruction. 208pp. 5⅜ x 8½. 0-486-46144-0

THE HUMAN FIGURE IN MOTION, Eadweard Muybridge. The 4,789 photographs in this definitive selection show the human figure — models almost all undraped — engaged in over 160 different types of action: running, climbing stairs, etc. 390pp. 7⅞ x 10⅝. 0-486-20204-6

THE IMPORTANCE OF BEING EARNEST, Oscar Wilde. Wilde's witty and buoyant comedy of manners, filled with some of literature's most famous epigrams, reprinted from an authoritative British edition. Considered Wilde's most perfect work. 64pp. 5³⁄₁₆ x 8¼. 0-486-26478-5

THE INFERNO, Dante Alighieri. Translated and with notes by Henry Wadsworth Longfellow. The first stop on Dante's famous journey from Hell to Purgatory to Paradise, this 14th-century allegorical poem blends vivid and shocking imagery with graceful lyricism. Translated by the beloved 19th-century poet, Henry Wadsworth Longfellow. 256pp. 5³⁄₁₆ x 8¼. 0-486-44288-8

JANE EYRE, Charlotte Brontë. Written in 1847, *Jane Eyre* tells the tale of an orphan girl's progress from the custody of cruel relatives to an oppressive boarding school and its culmination in a troubled career as a governess. 448pp. 5³⁄₁₆ x 8¼.
0-486-42449-9

JAPANESE WOODBLOCK FLOWER PRINTS, Tanigami Kônan. Extraordinary collection of Japanese woodblock prints by a well-known artist features 120 plates in brilliant color. Realistic images from a rare edition include daffodils, tulips, and other familiar and unusual flowers. 128pp. 11 x 8¼. 0-486-46442-3

JEWELRY MAKING AND DESIGN, Augustus F. Rose and Antonio Cirino. Professional secrets of jewelry making are revealed in a thorough, practical guide. Over 200 illustrations. 306pp. 5⅜ x 8½. 0-486-21750-7

JULIUS CAESAR, William Shakespeare. Great tragedy based on Plutarch's account of the lives of Brutus, Julius Caesar and Mark Antony. Evil plotting, ringing oratory, high tragedy with Shakespeare's incomparable insight, dramatic power. Explanatory footnotes. 96pp. 5³⁄₁₆ x 8¼. 0-486-26876-4

Browse over 9,000 books at www.doverpublications.com